Technology is Dead

How did we end up here, masters of scientific insight, purveyors of ever more powerful technologies, astride the burning planet that created us, and now responsible for cleaning up the mess and determining the future direction of all of life? And what do we do about it?

Technology is Dead is a book that attempts to answer both of those questions. It is a book of both challenge and hope, written for those who are able or willing to lead us out of our global predicament. It is a book for everybody: the politicians, CEOs, community leaders, everyday parents, and young people who understand that we must change our ways to ensure a sustainable future for all living things and the planet we rely on.

Christopher Colbert is the former Managing Director of the Harvard Innovation Labs and a global speaker and advisor on the convergence between technology, innovation, and humanity.

Technology is Dead

The Path to a More Human Future

Christopher Colbert

With Contributions from David Boghossian

CRC Press
Taylor & Francis Group
Boca Raton London New York

CRC Press is an imprint of the
Taylor & Francis Group, an **informa** business

Designed cover image: Guillermo Altube

First edition published 2025
by CRC Press
2385 NW Executive Center Drive, Suite 320, Boca Raton FL 33431

and by CRC Press
4 Park Square, Milton Park, Abingdon, Oxon, OX14 4RN

CRC Press is an imprint of Taylor & Francis Group, LLC

© 2025 Christopher Colbert

Library of Congress Cataloging-in-Publication Data
Names: Colbert, Christopher, author.
Title: Technology is dead : the path to a more human future / Christopher Colbert.
Description: First edition. | Boca Raton, FL : CRC Press, 2024. |
Includes bibliographical references and index. |
Summary: "How did we end up here, masters of scientific insight, purveyors of ever more powerful technologies, astride the burning planet that created us, and now responsible for cleaning up the mess and determining the future direction of all of life? And what do we do about it? Technology is Dead attempts to answer both of those questions. It is a book of both challenge and hope, written for those who are able or willing to lead us out of our global predicament. It is for the politicians, CEOs, community leaders, everyday parents and young people who understand that we must change our ways to ensure a sustainable future for all living things and the planet we rely on. The book's premise is that technology (like capitalism) has been an unprecedented force for prosperity, while at the same time bringing unrelenting, and unplanned downsides. Technology has insinuated itself into every nook and cranny of modern society, ignoring many of our human truths while preying on our vulnerabilities. It has resulted in both profound economic progress and a multitude of troubling unintended consequences, from deepening divides and loss of collective responsibility to a growing list of existential threats. The only viable response is to reconnect with our collaborative roots and undertake what the authors call a humanist revolution, a global effort to redefine human progress, rebuild our core systems to contribute to that progress, and reset our all too human behaviors and aspirations, while becoming the active, careful human-first stewards of technology itself. The revolution, guided by what the authors call a 21st Century Humanist Code asks all of us to work towards the world we want to live in – for each and every human to become the center and source of change in their lives, in their communities and the world beyond. The future of humankind and our planet depends on it"– Provided by publisher.
Identifiers: LCCN 2024005651 (print) | LCCN 2024005652 (ebook) |
ISBN 9780367546236 (hbk) (hbk) | ISBN 9780367546229 (pbk) (pbk) | ISBN 9781003089902 (ebk) (ebk)
Subjects: LCSH: Technology–Social aspects. | Technology–Philosophy. | Humanism. |
Technology and civilization. | Disruptive technologies. | Sustainable engineering.
Classification: LCC T14.5 .C65 2024 (print) | LCC T14.5 (ebook) | DDC 303.48/3–dc23/eng/20240408
LC record available at https://lccn.loc.gov/2024005651
LC ebook record available at https://lccn.loc.gov/2024005652

ISBN: 9780367546236 (hbk)
ISBN: 9780367546229 (pbk)
ISBN: 9781003089902 (ebk)

DOI: 10.1201/9781003089902

Typeset in Minion
by Newgen Publishing UK

"A great human revolution in just a single individual will help achieve a change in the destiny of a nation and further, will enable a change in the destiny of all humankind."

Daisaku Ikeda

Dedicated to my dear friend Alan Lewis—a fearless human who tried every day to be a better human and to help the world become a better place.

_____

Contents

Introduction

CONTRARY TO WHAT THE title might suggest, this is not a book about technology. This is a book about being more human. Technology, after all, in its seemingly infinite manifestations, remains a human project, responsible for many of the ways we experience the world—how we live, work, and exist. And yet technology itself is dead. It is inanimate, it is inhuman. It has no value until its capacities are *constructively* linked to the truth of our humanity and the positive, whole human outcomes we seek for ourselves, for future generations, and for the planet we rely on. Over the last 50 years we have allowed a troubling disconnection, a gap, to appear. Technology is now stewarding our lives instead of us stewarding it, resulting in a myriad of unintended, problematic, and even existential consequences. We must close the gap. And given the rapid advances of Artificial Intelligence, we don't have time to waste. We need to act now to establish a new order, one that puts human understanding and shared well-being first. My hope is that our collective effort will become a sort of Humanist Revolution, with millions standing up to fight; not with arms, but with persistent voices, shared sacrifices, courageous decisions, and model behaviors of what it means to be human.

In 2023 Adrienne LaFrance, the Executive Editor of *The Atlantic*, wrote an article titled "The Coming Humanist Renaissance" that underscored our task and opportunity:

> ...today we need a cultural and philosophical revolution of our own. This new movement should prioritize humans above machines and reimagine human relationships with nature and

DOI: 10.1201/9781003089902-1

with technology, while still advancing what technology can do at its best. Artificial intelligence will, unquestionably, help us make miraculous, lifesaving discoveries. The danger lies in outsourcing our humanity to this technology without discipline, especially as it eclipses us in apperception. We need a human renaissance in the age of intelligent machines.

Renaissance or revolution, Ms. LaFrance and I are both calling for radical change in how we interact with the world, technology, and each other. I propose the outcome we seek to be a shared 21st century humanist code, a universally agreed-to set of human first behaviors that becomes our species' guard rails and guiding light to achieve a more sustainable, equitable, and human future. *Technology is Dead* is about why we must do this work and how to do it.

The journey of writing *Technology is Dead* began while I was the Managing Director of the Harvard Innovation Labs, an incubator for Harvard student and alumni startups. While I was there I observed that, even though the founders were some of the brightest young minds in the world, 95% of their startups failed. And the primary reason for this was not their inability to develop their ideas but rather their inability to understand the humans involved: the customers, team members, investors, collaborators, in fact, all the humans, including themselves.

The Harvard entrepreneurs' presumption that technology trumped humanity resulted in failure after failure. I also came to realize that even when innovations, and particularly technological innovations succeed, they often produce unintended and costly societal, environmental, and human consequences. They either prey upon human vulnerabilities or ignore how inhuman we might become as technology increasingly defines our lives. The World Wide Web and its omnipresent poster child "Social Media" are obvious examples. Take a moment to think about what user-generated content, and its ability to be globally distributed at no cost and with no filters of truth have done to modern society: divisions, conspiracy theories, anti-vaccination movements, gun violence, drug epidemics, loneliness, and myriad other mental health issues. In a recent *Scientific American* story on the impact of social media on young people,

> Forty-one states and the District of Columbia filed lawsuits against Meta on Oct. 24, 2023, alleging that the company intentionally designed Facebook and Instagram with features that harm teens and young users.

Meta officials had internal research in March 2020 showing that Instagram—the social media platform most used by adolescents after TikTok—is harmful to teen girls' body image and well-being.

Many of the troubling modern realities we face as a species can at least be partially attributed to technologies being unleashed with no consideration of the downstream human effects. AI will no doubt be exponentially more consequential.

My summary plea: we need to spend less time writing technology code and far more time decoding humans, and then using that intimate understanding of us to make better decisions about pretty much everything beginning with how we guide and manage technology. The goal: more if not all of our so-called advances, decisions, and actions are net positive to our lives, the lives of future generations, and the life of the planet. Fundamentally, our existence must be centered on what it means to be human. This is not a new idea. Some of the greatest thinkers, from Petrarch and Erasmus to Bertrand Russell and Jean-Paul Sartre, made humanism their passion and purpose. There are countless organizations across the world, including Humanists International, currently doing the same. My task is to help create greater urgency for their work, offer a structured path for the actualization of our shared humanist aspirations, and enlist your support.

In August 2018 I was invited by the Monetary Authority of Singapore to be the keynote speaker at the Singapore Fintech Festival, a gathering of 60,000 technologists, business executives, bankers, and government officials from around the world. After much hesitation I agreed, with the qualifier that my talk could be titled "Technology is Dead." To their credit, the Monetary Authority and specifically Sop Mohanty, its Chief Fintech Officer, said yes. Three months later I was standing on a stage before several thousand people, calling on them to go forward not as bankers, businesspeople, or technologists, but as humanists. I asked them to put human progress and a more intimate understanding of our planet and each other in front of technology, and of every innovation we attempt and action we take. That clarion call would become the seedling of what I am now proposing: a global effort to re-balance our relationship with technology, to close the gap between us and it, and between each other. That moment made me realize that I am a humanist. It also prompted me to write this book.

I began writing in January 2020. Little did I know that a pandemic was lurking around the corner, a global conflagration that would magnify my point and the many issues we face. The almost 3-year battle to vanquish COVID-19 resulted in a groundswell of questioning regarding how the world works and how it should work to better steward our species and our planet. It prompted more people to challenge the conventional thinking that technology is the answer, and, instead, to propose that greater human understanding and elevation of its meaning is the only way forward. It was a transitional situation that humankind had experienced before. The pandemic was a lesser form of the bubonic plague, or Black Death, a ravaging of global civilization during the 14th century that wiped out an estimated 200 million people. Out of all the loss and misery the plague ushered in a new way of thinking and believing, an elevation of humanistic intentions, a re-birth. It caused a Renaissance. From human desperation came the motivation to examine our reality, re-define our intentions, and change our behaviors. In his article "How Pandemics Wreak Havoc—and Open Minds", *The New Yorker* contributor Lawrence Wright interviewed Gianna Pomata, a retired professor at the Institute of the History of Medicine at Johns Hopkins University, to explore the question of motivation derived from disaster.

Wright's interview excerpts below capture the clarifying and even restorative power of disasters.

> "After the Black Death, nothing was the same," Pomata said. "What I expect now is something as dramatic is going to happen, not so much in medicine but in economy and culture. Because of danger, there's this wonderful human response, which is to think in a new way…On the one hand, the plague works as a kind of acid. On the other hand, people try to re-create ties—and, perhaps, *better* ties.
>
> Like wars and depressions, a pandemic offers an X-ray of society, allowing us to see all the broken places. It was possible that Americans (the developed world) would do nothing about the fissures exposed by the pandemic: the racial inequities, the poisonous partisanship, the governmental incompetence, the disrespect for science, the loss of standing among nations, the fraying of community bonds. Then again, when people confront their failures, they have the opportunity to mend them."

The possibility of Pomata's "better ties" transpired for me during the writing of this book. In the bleak winter of 2021, I reconnected with Dave Boghossian, a colleague from my days at Harvard. A Harvard grad himself, a serial entrepreneur, and a fellow humanist, Dave and I began sharing our perspectives on what COVID was teaching us and what we aspired to for the world. He ended up becoming my thought partner and writing mate, helping add important dimension and clarity to my thinking.

Let me leave you with this. There is significant opportunity in the challenges we face as a species. An opportunity to re-think and re-focus on the world we really want. And then act accordingly. The thoughts, ideas, and potential solutions to follow are only a first step. And they are derived less from deep research and more a lifetime of observation, contemplation, and conversation about the human condition, forming a perspective that is no doubt skewed by my male, white, and Western upbringing. Know that my intent is not to prove my points beyond a shadow of a doubt but instead to raise awareness of our 21st century realities and motivate a movement that seeks to put our humanity at the front of every decision, at the front of technology instead of behind it. The goal is greater human intimacy and growth, of intellect, understanding, and deeper connection. And from that more well-being and a planet well cared for.

That movement is here, the Humanist Revolution has begun, and you're invited to join us.

I

The Technology Train

CHAPTER 1

How Technology Changed the World

"The great growling engine of change—technology."

Alvin Toffler, *Future Shock*

We are incapable of slowing our techno-centric world down. Technological innovation will continue to charge ahead, a runaway train fueled in large part by the realization of Moore's Law. The law is a widely validated theorem put forward by Gordon Moore the co-founder of Intel in 1965. Moore predicted that basic computing power would double every 2 years. And as it did, the cost to compute was cut in half. As computing power went up, and costs went down, life-altering change would happen. And it has. Since Moore first contemplated his theorem, technology has become a productivity lever in virtually every aspect of global society, to the point that technology appears to be everything and everything involves technology. The realization of Moore's Law has resulted in virtually every industry and individual on the planet being materially impacted by the ubiquitous ramifications of technology's bits and bytes. In his book, *Enlightenment Now*, Steven Pinker points out the significant progress we have collectively achieved on the back of technological breakthroughs. All the "functional" numbers are trending in the right direction, including how many transactions we can conduct daily, thanks to the fact that most of us in the developed world carry more computing power in our pockets than existed in earthly aggregate in 1965.

The statistics are truly compelling across sectors: cleaner water, greater access to health care, and declining poverty levels. According to the World

DOI: 10.1201/9781003089902-3

Bank, "Between 1990 and 2015 the global rate of extreme poverty fell at a rate of about 1 percentage point per year, from 36.2% to 10.1%, more than halving the number of poor over that period." More people are also living longer, and thanks to the Internet have access to more opportunities than they've ever had before. Interestingly, these improvements were not necessarily the intended target of most of the technological breakthroughs that helped produce them. Rather, I propose they were often derivative consequences of something else happening, of some other outcomes being sought and realized. And those outcomes and their unintended consequences, are worth delving into because in their examination potentially lies the initial clues to the three critical inter-related questions we must answer:

- How do we bridge the gap between the ever-accelerating advances of technology (the train) and humankind's ability to both catch up and one day steer future advances?
- How do we gauge and guide technological innovations to result in a collective upside while reducing the downside of unintended, negative consequences?
- How do we ensure that the intention of all technological breakthroughs and applications are the right intentions, including the rapidly advancing juggernaut called AI?

Technological innovations have always carried different goals and degrees of consequence, some desired and some not. They have had varying abilities to either improve existing systems or to disruptively replace them. Past innovations have presented themselves in a vast number of shapes and functions, from the wheel and written languages to robotics and the taming of fire. While they have come in many forms, most successful innovations have tended to increase humankind's power over our circumstance by delivering two functional benefits: increased speed or expanded duration, and in some cases both. Speed and duration have been and arguably still are the essential contributors to the historical measures of human progress: economic growth (productivity) and biological growth, also known as longevity. Whatever the origin, there appears to have been an implicit understanding among early innovators and innovation adopters that progress meant growth and that we can best achieve growth by either accelerating matters or by extended shelf lives. Since the invention of the first wheel, the innovations that have tended to get the most welcome receptions were the ones that did one or both of those things. In the implicit, amorphous quest to accelerate human progress, speed and duration arose as the surrogate and sufficient measures, both quantifiable and of immediate value to most. Faster and longer was, and to a large degree, still is perceived as better.

A QUICK HISTORY OF SPEED AND ELONGATING SHELF LIVES

Often cited as the first "technical" innovation, Johannes Gutenberg's invention of the printing press in 1440 produced the ability to print multiple copies of a written entity. What it also did was reduce the distance between questions and answers, between more people wanting to know and more people being able to know. The printing press helped people connect to understanding and to each other, and with greater connection greater collaboration ensued. The printing press effectively sped up our ability to encode, store, and share knowledge, which then sped up the development of new ideas that would, in turn, result in more and better ideas, greater growth and the sense of collective gain. The printing press sped up economic progress.

Electricity, heralded as one of the biggest breakthroughs of modern civilization, delivered more than light and energy. At the outset, electricity delivered longer days. It gave us more time to do more things. When you have more time to do things, you can accelerate all kinds of growth, from consumption to creation. Sanitation technologies, including boiling water, were another duration agent. They profoundly reduced the odds of infectious diseases and in doing so significantly increased the longevity of human life. The lifespan of the average American male in 1930 was 60 years. In 2019 it was 79. Joining the duration club: vaccines, refrigeration, the optical lens. Of note, according to the CDC, life expectancy in America in 2022 declined to 76 years, due in large part to COVID-19 fatalities, a pandemic indirectly fueled by technology. The duration agent became a double agent.

The combustion engine enabled things to be made faster for longer. Nitrogen fertilizer did the same. And so, did the Internet. Its invention in 1990 by Tim Berners-Lee was arguably the biggest technological breakthrough in all human existence because its seamless, connective capacity made everything faster and its duration potentially without limits or in some cases, irrelevant. When combined with lightning-fast microchips, laptops, and smartphones, the pace of the world and work went from being calculated in days, weeks, and months to being calculated in micro-seconds with no boundaries. Suddenly, we could trade, negotiate, buy, sell, barter, talk, listen, study, examine, calculate, and decide on any topic at anytime from anywhere. There were no limits on what could be done or how quickly we could do it. Our waking hours had been sped up; what we could do in one hour was seemingly multiplied by an order of magnitude or ten. And every industry was effectively sped up too. Age old slumbering businesses were rocked out of bed and tasked with re-engineering how they did what they did to get it all done much, much faster. The more recent poster children for technological advancement, Artificial Intelligence, Machine Learning, and the enabling of Big Data, are all effectively speed machines. They reduce the amount of time it takes to do things, solve problems, answer questions, create content, and at least theoretically make better decisions. The coming next generation, from quantum computing and 5G telecommunications to applied blockchain and edge computing are also fundamentally about making life happen faster or more efficiently, setting us up to have more time and, perhaps accidentally, feel greater pressure to do even more with that time.

A 21ST CENTURY ADDICTION

With time the central productivity variable, speed has become a modern addiction of sorts, one that has created a desire for more speed. Going faster has created an insatiable desire to go even faster and an unspoken expectation of 24/7 convenient access to anything: knowledge, capital, resources, and each other. As Columbia Law Professor Tim Wu pointed out in his 2018 *New York Times* article "The Tyranny of Convenience", speed, and its bedfellow, convenience, dominate modern society:

> In the developed nations of the 21st century, convenience — that is, more efficient and easier ways of doing personal tasks — has emerged as perhaps the most powerful force shaping our individual lives and our economies. This is particularly true in America, where, despite all the paeans to freedom and individuality, one sometimes wonders whether convenience is in fact the supreme value. As Evan Williams, a co-founder of Twitter, recently put it, 'Convenience decides everything.'

With immediate access to everything, duration as a measure of progress and value has taken a back seat. The longevity goal of innovations past has largely been replaced by a new one: the goal of moment-in-time relevance. Current technology has replaced the value of *longer* almost entirely with the value of *faster*, resulting in a focus on the present and a presumption that the future can and will be modified later. As much as duration and longevity carry an implied integrity of sustainable meaning, speed celebrates, motivates and rewards transaction-dominant, value-now human behavior. Our ability to forgo short-term tactical need for longer-term collective societal gain has diminished. And the same is also true of commerce. Industries, companies, brands, and products that have enjoyed multi-year, often multi-decade, shelf lives of relevance now face the daunting reality of a shelf life of zero. Access to anything has translated into the collapse of competitive barriers and the ability of anyone to create a better mousetrap with little cash and zero infrastructure. A long and storied corporate history has become of nominal worth and a going concern's imminent future is without guarantee. The average lifespan of an S&P 500 Index company in the 1950s was about 60 years. In 1965, 33 years. Today, it is down to about 20. A clear pattern is emerging (see Figure 1.1).

As the lifespan of companies has shortened so too has customer loyalty. The need and ability to move faster has fundamentally changed the nature of the relationship between us and what we buy, who we buy it from, and why. Whether a business has been in business for 100 days or 100 years, today's customers are remarkably open and receptive to even slightly

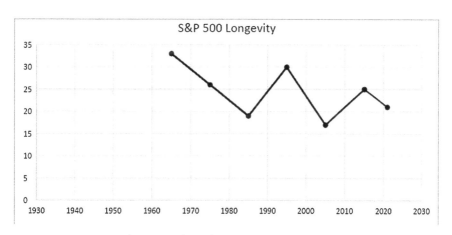

FIGURE 1.1   Average longevity (years) in S&P 500.

better or cheaper alternatives. The longstanding marketing tactic of multi-generational packaged goods companies featuring their "Founded" date on the label is no longer relevant to the consumer. Today is purely about "the here and now." The Internet has made comparison shopping as easy as one-click, and not just on price but on value, on peer reviews, and on delivery speed. But more importantly it has created a culture of disloyalty across all channels. We buy what's best in the moment versus what served us well yesterday. The world of commerce is now built for speed.

As brands are becoming disposable so too are most goods and services. In the clothing industry "Fast Fashion" has entered the picture, leveraging global supply chains, cheap labor, and cheaper materials to be able to offer the latest designer looks and styles at unfathomably low prices, enabling fashion forward shoppers to stay *au courant* at very little cost. Right before the pandemic, American consumers snatched up five times as many clothing items as they did in 1980, with the average top, bottom, or accessory being worn only seven times before it's thrown out. The technology-triggered tectonic shifts in supply side availability and buyer behavior have even rocked the CPG (Consumer Packaged Goods) brand world, resulting in a dog-eat-dog marketplace and a relentlessly commoditizing pressure that makes holding on to store shelves and market positions really tough. A 2015 study by Catalina, the company that got its start doing in-store couponing, revealed that of the top 100 CPG brands, 90 had experienced share declines in the last year. These are the all-American brands like Kraft mac and cheese, Coke, and Oscar Meyer hot dogs—all brands that built their businesses on the back of a tried and true formula: establish a solid value proposition for the product, promote it with explanatory advertising, price it right, and make it available everywhere. If you do this, market share will grow and stay. Well, that's no longer true. The speed scene has made the here and now the only thing that matters. Shorter shelf lives and speed-induced disposability has also come to furniture (IKEA anyone?), electronics, toys, and more. And if you add in digital formats, e.g., streaming music and downloadable books, it all translates into speed to buy, speed to use, and speed to move on.

THE WAR ON FRICTION

With speed the dominant measure of value, and time the most precious commodity, friction has become the common enemy—mechanical, economic, and social friction. Friction is the sticking point between where we are and where we want to be, it's the non-essential, non-value adding steps in the process. It's the circuitous mobile app interface that takes longer than necessary for no good reason, and it's the process bottlenecks that get in the way of getting it done. Friction is what slows us down. And in a world addicted to speed and enabled by quantum leap technologies, friction had to go. And so, it has.

Over the last 30 years as the world has become increasingly friction-less the ability to invent, to produce, to trade, and to share has become exponential. There have been massive and generally (but not always) positive shifts in how most (but not all) people live their lives. The centuries-old models of friction laden hierarchies and captive, plodding production systems that the user was subjected to are being replaced by systems and models that are frictionless, flat, fully distributed, and user empowered. And what's more interesting is that these new "open" standards demand that every entity, from country to company to city to you and me, needs to share those same characteristics. Closed and hierarchical top-down entities will ultimately fail or eventually be forced to opt-in. We need to accept the fundamental fact that what got us to where we are is unlikely to get us where we want to go.

There are hundreds if not thousands of other technology-enabled, speed-driven societal shifts that have transformed the world over the last 40 years. It strikes me that six carry much of the consequence: Individualism, Collectivism, Disintermediation, Value Networks, the Sharing Economy, and Democratization. Each has had a profound impact on how we live, work, and behave.

INDIVIDUALISM: POWER TO THE PEOPLE

Immediate access to information, resources, capital, and each other coupled with steady economic growth in much of the developed world has resulted in the rapid rise of Individualism over the least two decades. A 2017 research study led by psychology researcher Henri C. Santos at the University of Waterloo confirmed a global trend:

> Much of the research on the manifestation of rising individualism— showing, for example, increasing narcissism and higher divorce rates—has focused on the United States. Our findings show that this pattern also applies to other countries that are not Western or industrialized," says psychology researcher Henri C. Santos of

the University of Waterloo. "Although there are still cross-national differences in individualism-collectivism, the data indicate that, overall, most countries are moving towards greater individualism.

In America, the birth of Individualism can be attributed to its 18th century founding fathers and their focus on ensuring individual freedoms while limiting the arbitrary and potentially tyrannical powers of government that might get in the way of achievement. The critical qualifier was the presumption that individual freedoms would be tempered by the importance of the collective good, and that social institutions and systems would consistently reinforce that context. Robert Bellah and his co-authors of *The Good Society* cite the role of the English philosopher John Locke in framing that foundation:

> The Lockean ideal of the autonomous individual was, in the eighteenth century, embedded in a complex moral ecology that included family and church on the one hand and on the other a vigorous public sphere in which economic initiative, it was hoped, grew together with public spirit...The eighteenth-century idea of a public was... a discursive community capable of thinking about the public good.

In the 18th century, America freedom of speech was meant within the context of being able to speak against unfair, unjust rulers. Freedom to bear arms was meant as a protection against armed governmental overreach. What the forefathers could not predict was how time and technology would replace the context and the original intended meanings of freedom with adulterated, hyper-individualized ones. The presumption of the public understanding the importance of and being willing to contribute to the public good has been overwhelmed by the forces of self-interest and the presumed ability to exist alone.

As more and more citizens feel empowered to act and operate as individuals, they are also empowered to consume and communicate that way. In today's world, if you have the money, it is possible to get almost anything you want, when you want it, just the way you want it. And to have your voice heard by thousands and even millions of people at the push of a button. Bespoke sneakers, custom colors for your kitchen appliances, emojis that perfectly capture the uniqueness of you, almost everything in your life can be personalized to reflect your preferences, predilections, and personality. And it, whatever it is, can be delivered directly to you. Having it our way changes the game between society and ourselves, empowering and enabling us to switch careers on a dime, to book a vacation in a minute, to change banks with two clicks, or to declare that passion and purpose will now be our vocational guiding lights. Today's consumer has entirely different expectations than the consumer of not that long ago. There is a new set of rules, a new code between us and the world. The institutions that surround us are increasingly viewed as necessary evils, systems that we accommodate to get our way versus resources essential to elevating the collective whole. They are no longer in charge, we are.

We now expect to be in control of the decision from beginning to end. We expect to have infinite choices and the ability to dispose of things as readily as we buy them. We eschew long-term commitments and increasingly prioritize and pay a premium for experiences over products. In a time-starved life, with nanosecond access to anything, experience is increasingly what we seek, because it is where distinction and life-defining value appears to lie. Products are commodities. Increasingly we expect sustainability, because as much as we want to be able to throw things away, we increasingly want to know that the companies we buy from don't, reflecting the growing collision of sorts between the world of me and the world of we and the impact of the constant climate news. The two rules that likely carry the most weight are control and optionality, because with our current speed addicted, anti-friction mindset, we have no time to spare. Thanks to speed, Individualism is now in the fast lane and our collective democracy is behind the wheels of a tow truck.

The rapid rise of Individualism extends to the rapid-fire creation and dissemination of knowledge, our knowledge. The traditional media, once the voice of reason and intellectual truth, of professional and objective reporting, has been largely replaced by social media and the voices of the populace and an explicit new societal function called "Influencers." While influential people have always held sway over others, in today's infinitely connected world self-proclaimed Influencers are able to sway people at scale, which is to say, one Instagram post can make a stock go up or down, get a politician elected, or kill a new product launch. Individual influence, fueled by the ready access to infinite resources and others, and delivered via flat, open and frictionless systems, yields more resources, resulting in a looping power system that serves as an accelerant of sorts for those who are increasingly at the front of the room. According to Data Bridge Marketing Research, the global influencer marketing platform industry is projected to be a $70 billion business by 2029.

COLLECTIVISM: THE SEEDS OF TWO DIFFERENT REVOLUTIONS

As the loop loops and Individualism has emerged as the dominant modality of our daily existence, two different forms of collectivism appear to be emerging. One is the cast of nationalism and populism, less a matter of national heritage and values and more an angry stance against others who aren't like us, whether they're intellectuals or people from other countries. The other form is a growing collective of global activism and future concern. Arguably two sides of the same coin, both "causes" are fueled by the speed of everything and issues of economic inequality, immigration, and our ability to survive and combat existential uncertainties in the short- and long-term. It is the activist collective that represents our greater hope and is our greatest chance to bridge the gap between technology and us. The collective must coalesce around a common vision, a holistic measure of human progress, and be organized around common beliefs and behaviors. And it must work towards achieving a level of global collaboration and conformance needed to steer our shared planetary ship. That collective will become the Humanist Revolution.

DISINTERMEDIATION: THE DEMISE OF THE MIDDLE

The third big shift served up by technology is disintermediation, the removal within industries of layers of insufficient value creating functionality (aka friction) within their value chains. Disintermediation has impacted almost every sector, from travel and the now-dominant online travel booking engines to banking with peer-to-peer payment apps like VENMO and the promise of digital currencies. The entirely global financial system is frantically attempting to understand the true disintermediation capacities and downstream consequences of blockchain technologies and the associated DeFi movement. In a 2020 article titled "Is Disintermediation the Future of Finance" International Banker wrote,

> But in the wake of the 2008 financial crisis and with pertinent questions raised surrounding the efficiency (or inefficiency) of the global financial system, much of the focus has been on examining whether the existence of such intermediaries is always—or, indeed, ever—necessary. This has led to a strong trend towards disintermediation emerging over the last decade, with the process well underway across several distinct areas within the banking and finance sphere.

Regardless of industry or institution, eliminating middlemen players or steps has removed opacity, leveled playing fields and in most cases reduced both provider and end-user costs. It has exposed the hierarchy of industry for what it was, a Byzantine production system built when time was not so precious, duration was a given and friction was accepted as a cost of doing business. As the layers have been removed there has been a concomitant increase in Individualism as more and more of us expect what we want or need to be just right and right now. In response, Starbucks is opening mobile only stores, Amazon Go's retail experience allows you to check out without checking out, and UPS really is beginning to use drones for delivery.

VALUE NETWORKS: REMOVING WALLS

The fourth shift involves the bridging of industries and value chains into value networks. Until the 21st century, one of the main platitudes of many a business school professor was the avowed importance of never mixing business models, an action sure to result in abject commercial failure. The decades-old thinking was that commingling businesses with different value propositions, target audiences, and underlying growth engines would only produce market confusion and production inefficiencies while thwarting scale and profitable growth. The melded enterprise would surely fail. This was all true until about 10 years ago, when technology, speed, access, the decline of shelf lives, and the empowered buyer became the new norm. In today's technology-redefined world, combining business models results not in loss but in leverage. Singular value propositions are being replaced with symbiotic value networks. And while examples abound, Amazon may be the best.

Founded in 1994, Amazon spent much of its first 15 years building the largest retail e-commerce platform in the world, focused on selling consumer goods and merchandise. In 2002 they launched Amazon Web Services (AWS), the beginning of their cloud data management offering, and a seeming non sequitur to their core business. But not really. The e-commerce business supports millions of merchants who have trillions of data records that need to be stored somewhere. In 2022, AWS generated $80 billion in revenue and $23 billion in profit, roughly 90% of Amazon's total profit. Imagine a retailer 10 years ago deciding to get into the IT services business? Wall Street would frown.

Amazon is taking down more walls. While they are not building a bank, they are providing many of the more profitable services of a bank to their retailers, driving both profitability, but as importantly, stickiness for their members and merchants. The new order of open, flat, and distributed has profoundly changed the rules of the game, freeing up established companies and startups to completely re-imagine and re-build what it means to provide value at scale in today's world. Technology has blown away the preconceived notions, the do's and don'ts and the belief that rigidity and repeatability are the requisite attributes of any successful business.

THE SHARING ECONOMY: DON'T BUY, BORROW

Disintermediation, the success of value networks and individual empowerment all contributed to the fifth big shift, the Sharing Economy. It's a consequence of technology that's far more profound than it seems. Consider that most of us have grown up with very set beliefs regarding the importance of owning one's own assets and the idea that strangers are not to be trusted. Seamless, frictionless technology, ubiquitous availability, and self-managing validation systems coupled with the unquenchable thirst for *faster*, have resulted in new ways to assess trust and new motivations to let go of the need for ownership and control. As Richard Branson, the founder and CEO of Virgin once declared, "… we're seeing the sharing economy boom—I think our obsession with ownership is at a tipping point and the sharing economy is part of the antidote for that."

In a few short years the Sharing Economy has grown to trade in a vast range of products, functions, and services, from cars, co-working spaces and secondary market goods to clothes, accommodations, and crowd-funded capital. The benefits are equally diverse, including cost savings, convenience, control, choice, and flexibility. It's a multi-faceted, moment in time value that plays perfectly to the idea of the new world being open, flat, and distributed. No commitment is required, and you can buy, sell, rent, invest, or raise money, really, really fast. According to the accounting and consulting firm PwC the sharing economy is expected to reach $335 billion in revenue by 2025. By then over half of all Americans will be part of it.

DEMOCRATIZATION: ACCESS TO WHAT MATTERS

The final shift is democratization or the increased ability of more people across the planet to access humankind's three essential levers: capital, education, and health care. Much has been written about technology as a democratization engine and it is true. Consider how the combination of the

World Wide Web and smartphones have changed the lives of billions of people. Twenty years ago, a rural farmer in Kenya had no ability to transact business without paper currency and certainly no ability to obtain credit or even get to a bank. Launched in Kenya in 2007, the M-Pesa app converted a mobile phone account into a bank account, allowing its users to send, receive, and save money, all on their phone. It expedites (speed again) commerce, enables payroll distribution and tax collection, and reduces corruption and crime. Today, thanks to a remarkably simple phone app 37 million people across Africa, the Middle East and parts of Europe can bank without a bank.

Democratization is also slowing coming to education. Education continues to be a stratified offering, with 10% of young adults in the developed world never making it out of high school and, depending on the country, only 7 to 60% of high school graduates going on to college. In the United States just 40% of students graduate from college within 4 years and most of the rest never do. In the developing world the numbers are much starker, with 90% of students never making it out of high school. In Ghana, as an example, 50% of the population never gets past the 5th grade and very few Ghanaian teens graduate from high school (see Figure 1.2). Developed or developing, the primary reason for education limitations and attrition revolves around the challenges of physical access and/or the direct and indirect costs associated with being in a classroom. Thank to technology, that limiting reality is beginning to be turned upside down.

Early in the pandemic 1.2 billion children around the world were able to keep learning during the quarantine shut down because of online learning platforms (and their incredible committed teachers). They went from classroom to living room learning in a matter of days. Speed again.

The promise of online learning as a democratizing accelerant is real and in sight. The challenge is that while it significantly expands access, it produces another form of divide, the need for reliable Internet access and a computer. According to a study by the Office of Economic Co-operation and Development (OECD) there are vast access gaps between developed and developing countries and even in the United States, where as many as 25% of students from disadvantaged backgrounds do not have access to a computer. Clearly this is another gap between our humanity and the technology that leads us.

The final democratization lever is universal access to health care, and like education, it is a somewhat positive but uneven story. Technological

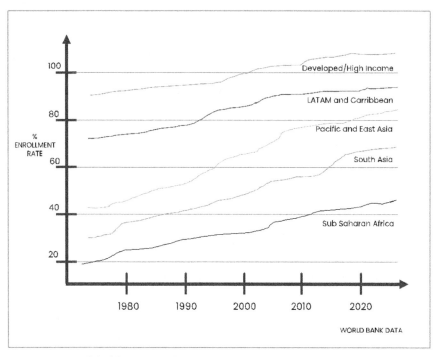

FIGURE 1.2   Global high school enrollment rates by region.

advances in the developed world have had a profound impact on mortality rates associated with certain diseases, have mitigated the most troubling conditions common with chronic illnesses, and simply have made basic health care more easily accessible to more people. As a recent example, thanks in part to COVID-19's quarantine requirements, tele-medicine became an effective and accepted standard of care almost overnight. It cuts a massive amount of time out of the equation for both the patient and the practitioner. Democratization of health care is also happening at an infra-structural level. A paper published by Stanford University calls out the big three in HealthCare IT: Intelligent Computing, Data Sharing, and Security, Privacy and Safety and how the technologies that underpin each are resulting in precedent-shattering levels of targeting therapies, data aggre-gation and analysis, treatment standardization, and the steady opening up of health information management and access.

The picture in the developing world however is not as positive. While more people today can access basic health services, 3.5 billion, half the world's population, still cannot. The progress made in the areas of HIV treatment, anti-malaria prevention, family planning, and immunization are due in part to technology but also to the incredible commitment of NGOs and philanthropic organizations like the Gates Foundation. Without their effort, the democratization of health care effort would falter and yet another gap would grow even wider.

THE TRAJECTORY OF TECHNOLOGY

How do the democratization of education, health care, and banking, coupled with the growth of Individualism, Collectivism, Disintermediation, Value Networks, and the Sharing Economy, add up to a better world for more people? The combined benefit, the outcomes intended or not, are more lives that are, on the face of it, more within our control, more accessible, offering more choice and providing more resources to more people. And the inferred consequence of that is better lives for now, and maybe even far better in the future given the trajectory of technology.

So, what is that trajectory? Beyond the now constant refrain regarding the impact of AI and Machine Learning (ML), and even quantum computing, what is the next generation of bold technologies and how does it both portend a better future for humankind and ideally provide bridges between it and us, effectively helping us close the gap and catch up with the technology train? There is no dearth of predictions and prognostications regarding what's next, but the one resource that rises to the top is the *MIT Technology Review*, the media platform of MIT, one of the most revered technology and engineering universities in the world. Each year MIT publishes its predictions for the next roster of most innovative technologies. The 2023 list, called "10 Breakthrough Technologies 2023" is as follows:

1. CRISPR for high cholesterol
2. AI that makes images
3. A chip design that changes everything
4. Mass market military drones
5. Abortion pills via tele-medicine
6. Organs on demand
7. The inevitable EV
8. James Webb space telescope
9. Ancient DNA analysis
10. Battery recycling

The question this list raises is not whether these predictions are right or wrong, but rather what they point to as intentions or desired human outcomes? What will these technologies result in and will they help to bridge or widen the gap between technology and us? A quick scan would suggest the following: innovators and technologists remain oriented towards increasing speed and duration with a growing concern regarding the environment, safety, and security. The latter makes good sense. The faster things move the more vulnerability can result, and vulnerability is in a way the gap between technology and us. The MIT list makes clear that the definition of progress may still be muddled by an addiction to faster and longer and that the question of how to bring more humanity into what we are inventing needs much more attention. How we communicate, what we value, what we seek, how we want to live, how we work, what we expect, our physical health, our mental health… all of it has changed due to technology. And while much of the transformation has been positive, it has exposed a gap between what technology is doing, and what we really want and need it to do. It is only by first understanding the gap and then intentionally working to close it, that we can change the trajectory we are on.

The train will never slow.

CHAPTER 2

The Growing Gap Between Us and It

"The saddest aspect of life right now is that science gathers knowledge faster than society gathers wisdom."

Isaac Asimov

We all know that the world is changing. That's not the problem. It's that we don't get just how fast it's changing, and how that technologically driven change is impacting our systems and ways of being that we stubbornly still assume are sufficient or sustainable. As a result, we don't take the need for our personal or collective change seriously enough. The bottom line is that if we ourselves don't evolve along with the systems and institutions that support us, we will go the way of the dodos and the dinosaurs. And there's another more implicit truth at play that is creating a massive gap between the pace of technological change and humanity's ability to keep up: it is much easier to change a discrete thing than it is to change the ways of us. A few years ago, the Santa Fe Institute, a research and education center focused on systems science, hosted a conference to explore what was titled "The Growing Gap Between the Physical and Social Technologies." As captured in one of the conference's workshop descriptions:

> ...the central hypothesis of the workshop is that the widening gap between technological advancements and lagging cultural and social structures and institutions is causing a variety of complex societal stresses and problems. These include concerns about rising economic inequality, fears over job losses due to automation, the increasing power of digital monopolies, the rise of populism,

DOI: 10.1201/9781003089902-4

growing criticism of democratic governance systems, loss of privacy and freedom, intensifying societal polarization, a loss of faith in experts and data sources, and growing dysfunction in key institutions. Looking further ahead, emerging technologies raise questions about what it means to be human itself.

Questions indeed. The gap between technology and us is becoming a gaping hole, a chasm perfectly, if qualitatively, depicted in Thomas Friedman's book *Thank You for Being Late* and conceptually visualized in Figure 2.1.

As the gap widens, the unintended negative consequences are stacking up. The core systems and institutions that underpin the way our world works are being exposed for being out of step with the demands of today (and tomorrow) and the humans trying to survive within it. Increasingly, many of our educational, health care, financial, economic, political, and governing systems are being called out for being behind. And the criticism is valid. These systems have become calcified and even paralyzed by legacy beliefs, out of date protocols, and behavioral sclerosis. The cross-sector

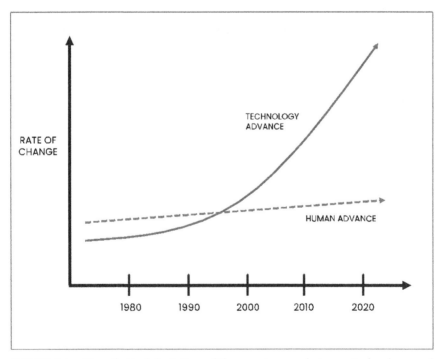

FIGURE 2.1 Hypothetical depiction of the growing gap between technology and humanity.

institutional struggle to change and to adapt, underscores a universal human behavior and bias: the rejection of the possibility that the way we are, the way it is, is not good enough. That all too human behavior is the biggest barrier to countries, companies, and individuals, including you and me, embracing the need to adapt faster, to close the gap and attempt to steer the technology train.

When technological change gets too far ahead of, or out of synch with, society's prevailing operating methods and humankind's ability to manage it in a healthy fashion, the unintended consequences proliferate. Many of those consequences played out in real-time as COVID-19 ran its course. The pandemic exposed matters of racial inequality, polarization, and inept, debilitated governance. It painfully magnified the economic divide with white collar workers holed up in their homes, being paid via direct deposit, while blue collar workers, and often people of color and minorities, were deemed "essential workers" and were required to risk infection and potential death to perform their low wage jobs. In the United States, and in many other "developed" countries, the pandemic revealed an epidemic of social and structural divide and decay. It turns out "developed" nation status is not so developed. The Santa Fe Institute workshop description was exactly right. Every troubling condition catalogued in that workshop description is unfolding in plain, painful sight today. And much of it has to do with the gap between the technology and humanity's ability to adapt fast enough. And our ongoing denial of the gap itself. Our work must start there.

SHOULD BLACK BOXES SCARE US?

This chasm between new technologies and existing societal, organizational, and human capacities can be attributed to one fundamental thing: we humans are much better at evolving discrete things than ourselves, and the web of systems and procedures we've created to get things done. The biggest contributing factor is human behavior, our reluctance to change. The second is sheer complexity. We have accidentally created a can of worms that can't be sorted out. There is an emerging view that our systems (local, national, and global) have simply outgrown our capacity to really understand them or to clearly see how changing one policy or component might impact other policies or components. In Harari's book *Homo Deus* he calls out the accelerating pace of change, the complexities and the confusion:

> When people realize how fast we are rushing towards the great unknown, and that they cannot count on even death to shield them

from it, their reaction is to hope that somebody will hit the brakes and slow us down. But we cannot hit the brakes for several reasons.

Firstly, nobody knows where the brakes are. While some experts are familiar with developments in one field… no one is an expert on everything. No one is therefore capable of connecting all the dots and seeing the full picture… Nobody can absorb all the latest scientific discoveries, nobody can predict how the global economy will look in ten years, and nobody has a clue where we are heading in such a rush. Since no one understands the system anymore, no one can stop it.

The gold rush surrounding Artificial Intelligence has only added to Harari's depicted incomprehensibility and its incumbent complexity and confusion. There is a proliferation of "black box" systems and algorithms, where humans acknowledge their inability to understand the logic of the machines, the predictions they are producing or how to manage them. That includes Sam Altman, the CEO of OpenAI, the creator of ChatGPT, who was quoted as saying "Society, I think, has a limited amount of time to figure out how to react to that, how to regulate that, how to handle it." Managerial questions and ethical flags are flying, and the clock is ticking. More and more data scientists are calling time out on the black box model, as reflected by a competition in 2018 organized by Google and academics from a range of prestigious universities titled "The Explainable Machine Learning Challenge." The intent of the competition was to prove the preferability of understanding ML predictions and their consequences versus simply accepting that the system "must be right." In an associated article for the Harvard Data Science Review on the competition, Duke Professor of Data Science Cynthia Rudin and Yale Associate Professor Joanna Radin concluded:

> The false dichotomy between the accurate black box and the not-so accurate transparent model has gone too far. When hundreds of leading scientists and financial company executives are misled by this dichotomy, imagine how the rest of the world might be fooled as well. The implications are profound: it affects the functioning of our criminal justice system, our financial systems, our healthcare systems, and many other areas. Let us insist that we do not use black box machine learning models for high-stakes decisions unless no interpretable model can be constructed that achieves the

same level of accuracy. It is possible that an interpretable model can always be constructed—we just have not been trying. Perhaps if we did, we would never use black boxes for these high-stakes decisions at all.

The emergence of black boxes and the seemingly impenetrable opacity of systems is vaguely reminiscent of Y2K, the code rush at the turn of the 21st century to make sure that all the information systems in the world would not collapse when the clocks struck 12:01am on January 1, 2000. There was a widespread fear that the time clocks in every computing machine would not be able to deal with the year value being changed from 19__ to 2___ and that the entire global infrastructure would crash. Well, it didn't. But what the massive $700 billion global fix-it frenzy did reveal is that no one really understood how these systems worked to begin with. The "systems" were just too complicated in part because they had been built over time, paint on paint layers of code attached to new code and a maze of patches along the way. There was no master plan, no blueprint, no documentation. Each "system" was just a mind-numbing tangle of files, folders, applications, protocols, and prior agreements that somehow magically delivered whatever the amalgam was supposed to deliver, sort of. The health care system in the United States is the poster child for this unrationalized complexity, this ungainly reality. We all agree that it does not work very well. We cannot agree on how to fix it because the fix is so incredibly complicated and because the system was never designed to do what it is now tasked with doing. American education is no less a can of worms, as are our economic, political, and financial systems. At a recent dinner party, I asked my companions to name the highest performing, most technologically advanced, and human considerate system in the United States today. The only answer anybody could come up with was Amazon's ability to consistently deliver pretty much anything within days if not hours. No other systems came to mind.

CLOSING THE GAP BY LEARNING TO STEER

In 2018 I spoke at a conference at Harvard University with Secretary Ash Carter, the Secretary of Defense under Barack Obama. We were jointly hosting an event titled *Technology Innovation and Public Purpose*. In his keynote address, Secretary Carter said this, "The pace of technological change cannot be slowed but it can, in fact, it must be steered." It must be steered. In his address at a Data Privacy conference in Belgium in 2018, Apple CEO Tim Cook echoed Carter's clarion cry:

> At Apple, we are optimistic about technology's awesome potential for good. But we know that it won't happen on its own. Every day, we work to infuse the devices we make with the humanity that makes us. As I've said before, technology is capable of doing great things. But it doesn't want to do great things. It doesn't want anything. That part takes all of us...we must act to ensure that technology is designed and developed to serve humankind, and not the other way around.

The "we" in Tim's and Ash's declarations is not the government or Apple, it is all of us.

As nascent revolutionaries we can begin by creating a more expansive, meaningful definition for human progress as our destination, and steer technology towards it, versus continuing to allow technology to steer us down a blind and non-humanistic alley. Along the way we will need to embrace the critical importance of, more intimately understanding humanity, beginning with ourselves, and putting that understanding at the center of our actions, decision-making, policy formation and support. And perhaps most radically we should be willing to help steer the institutions and systems that we rely on, demanding contextual and functional transformations that point towards a more sustainable, equitable, and human-first future. In their prescient 1991 book *The Good Society* the sociologist Robert Bellah and his co-authors spelled out that need, in fact, imperative:

> We badly need to enrich the way we understand our public institutions and comport ourselves regarding them, particularly by attending to how they affect or even create our identities as selves and as citizens. In an age of cynicism and privatized withdrawal, it may seem quixotic to call for a reinvigoration of an enlightened public. But we believe this reinvigoration is not an idealistic whim but the only realistic basis on which we can move ahead as a free people.

The authors' main points are central to our ability to close the gap. We should demand and invest in the re-design of our core institutions and systems as human-centric propositions that are laser focused on contributing in whole person ways to the collective effort known as human progress; and in order to do that, more of us must get intellectually and actively involved in moving those systems forward. Most fundamentally Bellah and his collaborators are saying that we need to reinvigorate our intellectual and activist engagement with the world if we want the world to be different and the trajectory we are on to deliver us to a better, more human place. It makes sense. We just need to do it.

A GREATER MOTIVATION FOR CHANGE

The in-our-face consequences of the COVID-19 pandemic and the quarantining of almost three billion people ignited many conversations about the weaknesses in our global economic, social, and democratic systems. These conversations were and are fundamentally about what it means to be human and how we want to live when we get to the other side, when we get to the future. The issues we are seeing right now in the developed world we have always seen. We have known about the divisions and inequalities, we have experienced the dysfunction and breakdowns of our society's core systems for years, and we have talked *ad nauseam* about the disconnectedness and isolation experienced as part of a 24/7 connected, technology-driven world. What was different in the unprecedented pandemic moment was how it felt, and in that feeling there now appears an almost painful clarity and perhaps more intimate understanding of human truths and consequences. The sudden loss of activity, of work, of the ability to socialize freely, coupled with the constant flood of imagery and stats capturing the staggering loss of life and the thousands of first responders and health care workers who risked theirs every day brought with it a newfound appreciation and a recognition that maybe, just maybe, the way life was, the way life is, is simply not the way we want it to be.

With those emerging questions comes a growing consensus that for all the wonders and improvements that technological innovation has brought into our lives, it's clear that it has also brought setbacks, separation, and even a loss of understanding of something fundamental, something about what it means to be human. My pushback is not, nor should it be, on technological progress but rather on humanity and whether we are willing to step forward to steer the train. Steering means working harder to better understand the unintended consequences that technology has spawned and to better understand the human behaviors it has perhaps unintentionally exploited and preyed upon. And to then use that intimate understanding to create guard rails and frameworks that can guide all future technological innovations and mitigate the negative effects on our society and planet.

Outside of nature, humans are the builders and/or destroyers of all that is. We must own that. As we gain greater understanding of how we are and what we (should) collectively seek, we will be far better able to articulate a desired future for humankind and begin to make the building decisions that will guide us, technology, and our institutions towards that goal. And we should be able to make decisions that are fundamentally less destructive to ourselves and our planet. The future of the world and our ability to improve its collective health, to regain our human footing, is directly tied to our ability to bridge the gap between what technology can do and what we must do to realize its best, most human potential. One of the greatest ironies of modern humankind is that for all our advances we appear to have already lost or are losing connection with our humanity; that for all our gains in scientific understanding, we appear to have regressed in our understanding of us. It's my belief that reversing that equation will become our greatest innovation, perhaps even our salvation. A deeper decoding of us is an essential next step.

We are human after all.

CHAPTER 3

Decoding Humans

"It has become appallingly obvious that our technology has exceeded our humanity."

Albert Einstein

As we seek to better steer future technologies and close the gap between us and it, we must first examine what it means to be human, from our mindsets and behaviors to our needs, desires, and vulnerabilities. A deeper understanding of us is key to how we manage our planet and the technologies that increasingly drive or influence our actions. It is the only way to ensure that technology serves us, versus us serving it.

All the coding in the world will not work if the coders do not first understand the human code. But decoding humans is incredibly hard. The complexity of humanity far exceeds the complexity of technology. Why humans do what we do is complex. Why we don't do what we should do is complex. Why we do what we should not do is complex. Why we are not inclined to dig into that complexity to increase the chance of our efforts being successful is even more complicated. It likely starts with our subconscious need to avoid discomfort. To study humankind is to study ourselves and that contemplation will no doubt result in feelings of fearfulness, fixations on control, social addictions, and a slew of other persistent primal and problematic needs. The stark realization about them becomes a painful realization about us. It is also a realization that cannot be easily synthesized, catalogued, or tapped into. The forensic examination of us, and the outcome of that examination, is messy. To open that can of worms is to enter a world of innermost truths and often uncomfortable admissions of want,

DOI: 10.1201/9781003089902-5

need, desire, greed, and lust. Such truths are both deep-seated and shallow impetus behind our actions and inactions. And they are the primary determinants of our willingness to embrace change or not. The combined conclusion of all this is that the real reason leaders and innovators don't seek to understand their customer, to decode their customer's humanity, may be because they don't really want to.

THE CODE ACCORDING TO MASLOW

And yet there is so much at stake. Millions of lives, thousands of good ideas, and trillions of invested dollars, and the future of Mother Earth could and would be better realized with a unilateral commitment to cracking the human code. With that cracking, the gap between us and the technologies that are racing ahead of us could be narrowed and perhaps eventually closed. And while it's a complex undertaking, we and they, the innovators, the social scientists, the public and private sector global and civic leaders don't have to solve every problem for every individual. We just must work to clarify our collective intentions, reset the context and function of the supporting institutions and systems, and provide our fellow citizens with the capacities and freedoms they need to solve their own challenges. And we can begin our decoding efforts by tapping into all of the clarifying psychological and behavioral research done since 1879, the year Professor Wilhelm Wundt established the first laboratory for the study of psychology at the University of Leipzig in Germany. Amidst the reams of scientific theory, one of the most profoundly simple and actionable perspectives on human behavior belonged to an American psychologist and university professor from the 1950s by the name of Abraham Maslow. A self-declared humanist psychologist, Maslow spent most of his academic career and research delving into the motivating forces and need states of humankind. Much of his original thinking culminated in a framework now called Maslow's Hierarchy of Needs, a theoretical construct depicting the five fundamental stages of need that most humans can or will go through during their life (see Figure 3.1).

Beyond the framework itself, Maslow's most profound insight was that humankind largely lives at the base levels of need. He observed that most of our actions and decisions, including whether to embrace a new something or somebody, are filtered through incredibly primal, survival-level needs. Our subconscious reptilian focus is always on ensuring our safety, comfort, and control to manage threats and more pointedly offset our nagging fear

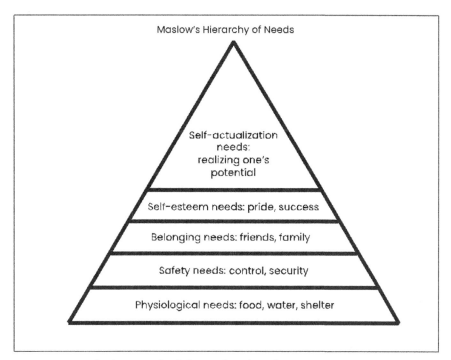

FIGURE 3.1   Abraham Maslow's hierarchy of needs.

of harm, loss, and even death. Thousands upon thousands of evolutionary years later we continue to fear the wooly mammoths, the saber-toothed tigers, and strangers. We fear people from different backgrounds, different lands, people of different colors and beliefs, because difference is outside of our control. Difference is a threat. The subjugation of the different by the dominant has been a constant in the history of the world. In America in recent years that inhuman human dynamic has once again become a raging sea of riots and protests, triggered by racist-laden declarations of wrongdoing, shocking episodes of police brutality, and an irrational fear of immigrants. This disturbing, unacceptable time in our so-called modern society is made even more disturbing by the fact it is a centuries-old, painfully repeating truth: a large percentage of our country, once considered one of the most progressive in the world, still thinks less of, and wants to suppress and even hurt, people who aren't like them and don't look like them. These groups are threatened by strangers and are desperate to feel in control. Sadly, it is not a uniquely American dynamic.

Technology has contributed in significant and positive ways to helping us meet Maslow's first level of need, enabling more people to have greater access to food, water, and shelter. And it has done a decent job of improving safety in multiple forms. It's when technology arrives at the third level need, Belonging, where its impact shifts from mostly positive to a mixed bag. As much as we fear strangers, we often, desperately, subconsciously, also seek their validation. Sean Parker, one of the co-founders of Facebook, confirmed that fact when explaining the social media company's multi-billion-dollar business success, "It's a social-validation feedback loop … exactly the kind of thing that a hacker like myself would come up with, because you're exploiting a vulnerability in human psychology." Technology allows humans to simulate connection, to fake both caring and being cared for. Our need for belonging is being met, but not really. And with that false enablement comes a myriad of unintended social and mental health consequences. At the fourth Maslovian level, the need for Self-esteem, the sometimes-negative consequences of technology become more serious. As technology has fueled the economic divide (as of today .01% of all humans control 11% of the total wealth in the world), it has also fueled an esteem divide. Our constant, unrelenting exposure to those who are revered, celebrated, and seemingly adored can often reinforce that we are not in their league.

At the fifth and final level of need, Self-actualization, technology may find itself less a friend and more a frequent combatant. Self-actualization is a wholly human, hyper-personalized undertaking that requires at least some freedom from the insatiable shackles of technology-enabled trans-action frequency and speed. Technology's capacity to expedite is a double-edged sword on the self-actualization journey, presenting both an efficient means to personal achievement and yet a constant distraction from the task at hand, which in large part is about letting go of what other people think to realize one's true potential. The idea of human progress could and should be viewed as collective self-actualization. And we should recognize that as good as technology is at providing food, water, and shelter, it needs to be far more carefully stewarded to be a positive contributor to our more aspirational shared human outcomes.

HUMAN COSTS VERSUS HUMAN BENEFITS

This direct contradiction between our primal fear of strangers and our primal need for their acceptance underscores the complexity of the human code and our confused quest for what we really want. A human's decision to buy or not to buy, to do or not to do, to embrace change or not, is both a subconscious and conscious return on investment calculation, incorporating deeply seeded needs, desires, wants, and fears. In our decision-making process we conduct both a back of the envelope calculation of whether the benefits sufficiently exceed the costs, and as importantly, a gut check. And in our investment calculations there are multiple kinds of benefits and multiple kinds of costs to weigh. Time, money, and function are the standard variables in the equation, playing roles as both benefits and costs. Investing time at work is a cost, reaping time in the form of a paid vacation is a benefit. Investing money is a cost, generating a positive return is a benefit. The function of cutting one's own hair during the pandemic was a cost, saving fifty dollars was a benefit. As central as these three variables are to how people make decisions their relative weight is often superseded by other factors, the <u>human</u> benefits and costs (often subconscious) that accrue to Maslow's Hierarchy of Needs. It turns out that when we are weighing the value of an act or a decision, we are weighing its impact on us in very fundamental psychological and physiological terms.

We saw this human calculus play out at the tail end of the pandemic when the world decided to open up again, after many months of lock down and shelter-in-place orders. There was a conscious and subconscious rolling calculation that went something like this: in the beginning the benefit of reduced physical harm, of death, overruled the likely cost of economic loss. Physical survival was more important than economic survival. The decision, led by our leaders and the epidemiologists, was social distancing, quarantine, and as a result, economic shut down. But over time the importance of the variables shifted, and new human ones entered the equation, particularly our base level needs for comfort, for control, and for the feeling of belonging. The voices of the epidemiologists faded. After a few months of mental and economic suffering we decided the following:

1. the risk of physical loss was less important;
2. the slow decline in fatalities was a sure sign of the problem being solved;
3. and that loosening the lock down was suddenly the right strategy; even though it would no doubt result in the same (or greater) surge in fatalities and demands on the already overwhelmed hospitals that originally motivated every country in the world to quarantine its citizens.

Suddenly the human benefit of freedom outweighed the human cost of physical harm. These shifting forms of logic reveal not just the irrationality of our collective decision-making but the over-weighted importance of human benefits and human costs in our calculations, as the leading force in both very big and very little decisions. Virtually every decision we make is impacted by our subconscious and conscious calculation of how a Yes or No will impact our survival, our safety, and our status. But there is another set of factors, the intermittent pull of something biblically catalogued as the Seven Deadly Sins. Effectively the dark side of Maslow's need pyramid, the seven sins, Pride, Envy, Gluttony, Greed, Lust, Sloth, and Wrath, all play a role to different degrees and at different moments in our lives and decision-making. Few if any humans are fully exempt from their distracting and unhealthy seductions, and fewer still are exempt from the never-ending quest to feed our basic needs. The vast range of costs and benefits and the omnipresent potential pull of the sins, all make for a complicated decision-making equation made even more complicated by another human truth: all major decisions are made emotionally. Human decision-making regarding the big decisions in life, from where to attend college and whether (and who) to marry, to what career to follow and which house to buy, are riddled with emotion and remarkably lacking in any analytic rigor to guide them. Some of the motivations are primal, others are inherited, and some are learned. Regardless of the source, our decision-making process, our code, is fraught with subjectivity and complexity. If we want to understand what to allow technology to do and not do, what we will do with technology, and how to ensure that technology does not hurt us, we really need to crack this code.

DECODING UBER

Uber and the other car-sharing startups in general are a case study in the importance of situational code cracking, the categorical understanding of the human benefits, human costs, emotions, and perhaps a sin or two at play. Inarguably Uber's pre-pandemic, pre-IPO valuation of $120 billion dollars was not based on its profitability since it hadn't any. Nor was it based on its intellectual property since it had none. And it's unlikely that the sky-high valuation was because its car service was slightly cheaper than a cab. Arguably Uber's valuation was and is a function of the investors' collective and emotionally charged calculation that it delivers on every benefit variable in the return-on-investment equation: money, time, function, and most importantly, two other Maslovian physiological and psychological needs. The brilliance of Uber and its car-sharing competitors is that they feed our now age-old addiction for speed and access while giving us something else that we all desperately, if subconsciously, seek and that is control. While car sharing efficiently gets us from Point A to Point B, what it really does is put us in the proverbial driver's seat called control, via the now guaranteed ability to know when the car will come, what kind of car it will be, who will be driving it, what he or she look like, how well-rated the driver is, what route he or she will take to get to us, how much the trip will cost, how long it will take, which route the driver will follow, and when exactly we will arrive at our desired destination. With car sharing there are no questions or worries, including whether we need our wallet. In the urban markets around the world where Uber, Lyft, Careem, and others have been allowed to operate, and the taxi industry has not materially responded, it is being put out of business, not because taxis are more expensive but because, in comparison, the taxi experience makes us feel out of control.

The matter of control extends to the decisions made by the approximately 3.5 million Uber drivers around the world. As Uber's customers undertake a cost–benefit analysis of whether to use or not use Uber, the drivers all undertake a similar analysis of whether to drive or not drive. And Maslow's pyramid of needs was no doubt central to their decisions. Driving for Uber helps each driver acquire food and shelter, its flexibility represents control, and its driver coterie of sorts provides some sense of belonging. Whether it contributes to the higher-level needs of self-esteem and self-actualization, particularly given the issue of living wages, is an open question.

THE GRAVITY OF ADOPTION

There are millions of technology innovations that were or are seemingly good ideas but bad businesses because they fail to meet humans where humans are. A quick visit to the startup and corporate innovation grave-yard provides ample evidence. When innovations do fit the human truths, when they meet our Maslovian needs, when the human return on invest-ment is clear, they are adopted, and with adoption tends to come com-mercial (but not necessarily societal) success. Successful innovation is not only about the ability to build the function, to bring the idea to life. It is also about affecting adoption of the function, to bring the idea into the customer's life. Adoption is different from purchase, or lease, or subscrip-tion. Adoption means to bring into a relationship. And that is a very high bar to reach. Henry David Thoreau, the 19th century American philosopher once wrote, "The cost of a (new) thing is that amount of life which must be exchanged for it." When humans take a new thing into their lives or work, they are giving up part of their finite capacity to do so. It is a profound commitment and one that works both ways. A big part of the failure rate of good ideas is fueled by our unwillingness to give up part of our lives, to sacrifice, or perhaps more bluntly by our inability to do the life math correctly. As much as there are calculations occurring in the assessment of decisions to be made in general, and innovations to be adopted, most are calculations riddled with unhealthy short-term needs, imprudent wants and dozens of to-be-explored irrational biases that cause us to make bad decision after bad decision. Consider retirement planning in America. Nearly half of all families today have zero retirement savings as compared with 40 years ago when most employees had pensions provided by their employees. The difference between the two scenarios is who made the decision. In 1980, companies made the decision that the employee should save for retirement. In 2023, it's up to us, and many of us are deciding that what we want now is more important than what we might need in the future.

BIAS: A VIRUS IN OUR CODE

Inside our decision-making code is an insidious creature we all manifest throughout our daily lives known as bias, or more specifically cognitive bias. Cognitive bias reflects our systematic tendency to overrule logic and rationality, opting instead for our own personalized reality, one that suits us just fine. The nature and behaviors of most of humankind are guided not only by the facts but by our preconceptions and biases, a muddled combination of thoughts and feelings conjured up, whether as part of a millennia old practice of adaptation, or simply a function of who we are and our limited ability to process information and make methodical decisions. Our inability to take into account every consideration effectively requires a leap of faith of sorts. Sometimes big, sometimes small, but always a leap influenced by pre-set notions and tendencies.

Out of the growing psychological sea of biases has emerged something called heuristics, also colloquially referred to as rules of thumb, a decision-making technique that humans use to make snap decisions or find quick, reasonable solutions to complicated problems. More often than not heuristics result in sub-optimal choices. One perspective is that heuristics came about as a desperate act of survival. Another is that it is derived from intellectual laziness. Regardless of the source, heuristics are our way of avoiding thoughtful, comprehensive, risk-adjusted planning and decision-making whether for our lives or the future of our planet. We opt for the here and now because it's simply faster (speed again) and easier.

Heuristics and biases get in the way of good decisions about pretty much everything and there are scores of bias types, as many as 188 according to the codex designed by John Manoogian III (see Figure 3.2).

When the long list is examined through the lens of humankind's willingness to invest in bad ideas, there are several decision-making and behavioral biases that play an integral part, ranging from Confirmation Bias, the tendency to look for or interpret information that confirms our preconceptions, to the Backfire Effect, the remarkable ability of people to receive data that refutes what they believe but results in them believing it even more. Biases have had a heavy hand in elections, wars, and marriages. And they have a heavy hand in how Artificial Intelligence works. To have technology serve us versus us serve technology, we must figure out how to remove bias and two other not helpful tendencies from our decision-making and behavioral equation.

GRANDFATHERING AND POLE THINKING

The story of the evolution of modern civilization is also a story of grandfathering. Each generation has carried forward notions, beliefs, traditions, and ways of doing things from the past that it believes are the right way to do things in the present. Just like us, institutional systems carry forward input standards, output norms, rules, and logic that might have made sense back then but may not make sense now. Grandfathering in any form is a perfectly human behavior because it reflects our preference for the comfortably familiar, an unconditional respect for the past, and a delusion that the present is not that different from "back then." It is also profoundly easier to hold on to what has been handed down to us than to create anew. Such generation-to-generation grandfathering by us can at times be a strength, a form of natural selection adaptation, but it quickly becomes a weakness when the layered complexities have added up to an impenetrable system hairball that is incapable of significant and rapid change. The past ways of us simply don't serve the future ways we need to be. Penetrating and unwinding the knotty hairball demands courage.

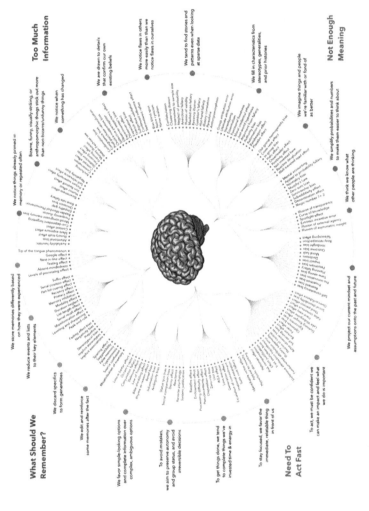

FIGURE 3.2 Cognitive bias codex.

It also requires an atypical level of critical thinking, of intellectual investment, and rare open mindedness that pushes our embedded biases out the door while welcoming in new perspectives. And it ultimately requires accepting the challenges of the transition from the way it has always been to the new way we want it to be. Again, it can be done, we just must try, and we must believe.

The complexity of human behavior and how it slows us down in the quest to catch up with technology change is also paradoxically manifest in our unhealthy and almost universal habit of looking for simple answers to complex questions. Another classic bit of human subterfuge can be seen at work when we tend to find what we want in the perceived purity of the extreme poles. Uncomfortable with nuance, details, shades of gray, and even compromise, we seek safety in the seemingly simpler world of black or white. That's why most humans like to make everything into a two-horse race: Conservative versus Liberal, Isolationist versus Globalist, God versus the Devil, East versus West, Woke versus Anti-Woke. The list of two choice categories goes on and on and on.

Pole thinking is inherently flawed because rarely, if ever, is anything that simple. Let's return to the health care picture for a moment. The truth about America's health care problem is that both partisan poles are wrong or at least not perfectly right. Even if we could agree that cost effective and caring health care should be available to all who need it, the current views on how to deliver that tend towards the almost opposite poles of private markets (demand and supply doing their thing) versus a government program (nationalized and largely paid for). One doesn't do enough, the other might cost too much. So, what's right? What's right is finding the middle, the complex and nuanced just right shade of gray that reflects essential compromise and a deep understanding by all the policy-creating participants of how the current system works. It requires a rigorous analysis of what the tradeoffs must be, what possible unintended consequences will arise, and how we might re-design things to ensure that our health care delivery improves across every facet of performance while not bankrupting the Federal Government. The problem with this approach is that it's complicated, aka hard. Our leaders don't seem to have the voter incentive, intellectual stamina, or rigor to do the work to get to that point of policy development proficiency, or the principles and grace to compromise. And we don't either. They and we jump to overly simplistic conclusions that reflect our overly simplistic black or white positions to satisfy our very individual need to just have an answer we can hold onto and that serves us, to remember and identify with, even if it's wrong. As we march forward in the Humanist Revolution the color of our flag will necessarily be gray.

SYSTEM 1 AND SYSTEM 2

The story of the growing gap between the pace of technological innovation and humankind's inability to keep up or catch up, is fundamentally a story of our thinking and feeling systems, the human code at work. In his seminal 2011 best seller *Thinking Fast and Slow*, Daniel Kahneman, a winner of the Nobel Memorial Prize in Economic Sciences, depicts two biologically built-in decision-making modalities that humans deploy to navigate life and determine how to respond to pretty much any situation. When faced with choice or challenge, humans seamlessly and subconsciously toggle between what Kahneman labels System 1 and System 2 thinking to arrive at an acceptable (to them) response. System 1 is lightning fast, emotionally charged and highly intuitive. System 2 is its plodding cousin, a logical, methodical and much slower means to arrive at an answer. System 2 tends to kick in to either double check the work of System 1 or to save the day when System 1 can't handle the complexity it is facing. Interestingly, neither system is exempt from the centripetal force of bias. In describing how the systems partner, Kahneman writes,

> System 1 continuously generates suggestions for System 2: impressions, intuitions, intentions, and feelings. If endorsed by System 2, impressions and intuitions turn into beliefs, and impulses turn into voluntary actions. When all goes smoothly, which is most of the time, System 2 adopts the suggestions of System 1 with little or no modification.

The operative phrase there is "most of the time." Some of the time, in decision situations laden with uncertainties, unknowns and daunting complexity, System 2 and its carrier, the human, should step up and overrule System 1. But it doesn't and we don't. Because it turns out that embedded in the systems are two other debilitating tendencies. We have a constant and raging desire to get to the answer as quickly as possible and we are unwilling to apply much intellectual rigor to the questions of why, whether, what, who, or how.

Kahneman characterizes this lack of rigor as "intellectual sloth" and explains,

> Those who avoid the sin of intellectual sloth could be called "engaged." They are more alert, more intellectually active, less willing to be satisfied with superficially attractive answers, more skeptical about their intuitions.

And simply better at making improved hard decisions. The current cultural decline of intellectual rigor and the rise of sloth, two sides of the same coin, appear to be societally and technologically fomented, opposing curves rapidly steepening over the last few decades, particularly in America but arguably throughout the world. Not just central contributors to suboptimal decisions in general, they are central factors in the issue of the growing gap and present an alarming barrier to our collective ability to make better decisions to benefit more of humanity, and to render material human progress. The adulation of speed as the primary measure of everything, the replacement of the Fourth Estate by user-generated content, and the view that adult study is an antiquated, irrelevant function of the past are just a few of the technology fed forces that have contributed to what could be called intellectual rigor mortis. The Western world is experiencing a slow, painful decline amid the growing erosion and derision of intellectualism and critical thinking, an insufficient understanding among most people of the way the world works, and a lack of interest in the topics and issues that really matter. A frightening finding is that, according to an Annenberg Constitution Day Civics Survey done in 2017, only 26% of Americans can name the three branches of government. A C-Span poll that same year revealed that 52% could identify one Supreme Court justice, and according to a 2016 National Science Foundation Survey one in four Americans think the sun orbits the earth. Yes, the earth.

REWRITING THE CODE

In that somewhat disheartening view lies hope and perhaps the only means to solve the puzzle:

- We must accept that there is no secret algorithm. That bridging the gap between technology and us, of transforming our systems and institutions to contribute to collective progress, of vanquishing the growing global threats, simply requires more of our citizenry, leaders and innovators becoming one of Kahneman's intellectually "engaged." Humanist Revolutionaries if you will.

- The ultimate solution to finding the answers lies in more of us being willing to dig deep into Maslow, the biases, the sins, the systems, and the psychology of human behavior in general. And to do so without fear, while embracing intellectual rigor as our core commitment, and becoming open to taking risks and making sacrifices for the good of the whole. That is revolutionary behavior.

As captured by Bellah and his co-authors of *The Good Society*, "Our primary task is to recover a public capable of understanding and so of enacting a genuinely democratic government." They go on to declare the importance of more of us embracing the responsibility for a common life, a better life for all, citing the then newly democratized Eastern European nations and a portion of Vaclav Havel's, the elected president of Czechoslovakia's, 1990 New Year's Day speech:

> Let us teach ourselves and others that politics (governing) should be an expression of the desire to contribute to the happiness of the community rather than of a need to cheat or rape the community. Let us teach ourselves and others that politics can be not only the art of the possible, especially if "the possible" includes the art of speculation, calculation, intrigue, secret deals, and pragmatic maneuvering, but it can also be the art of the impossible, that is, the art of improving ourselves and the world.

A decision to improve ourselves and the world is to break with the broken human decision-making conventions and challenge the current support structures. The revolution starts there. It is to risk being different, holding to the loneliness of independent conviction while accepting that compromise, the middle ground, is likely the best choice. Risk really begins with having the willingness to look in the mirror to examine ourselves, our behaviors, and how our own primal needs and biases are influencing our self-interest and the choices we are making. Our ability to understand other humans starts with understanding us, and sacrificing some of the things we want, perhaps the things we find comfort in, to forge new views and potential paths. It's an idea well conveyed in the words of Martin Luther King Jr.,

Human progress is neither automatic nor inevitable... Every step toward the goal of justice requires sacrifice, suffering, and struggle; the tireless exertions and passionate concern of dedicated individuals.

The capacity of any entity to overcome its enemies, whether internal or external, lies in the ability of its leadership and citizenry to do two things: take risks and make sacrifices. Giving up power, compromising our position, walking away from maximum profit, foregoing the convenience of water in plastic bottles, opting not to get on a plane, putting down the phone to talk to our children, all are sacrifices. And sacrifice is the bane of self-interest. Until we are personally impacted by the enemy itself, most of us will not make those sacrificial choices. We prefer to seduce ourselves into believing that holding on to what is, maintaining the status quo, waiting it out, and not personally giving up much of anything, will somehow magically turn into a turnaround. Or maybe we simply don't care about the turnaround because most of us will be in heaven or hell by the time the devastation arrives at our door. It's a mainstream societal and corporate approach that perfectly captures that classic definition of insanity: doing the exact same thing over and over and expecting a different outcome.

Sacrifice includes giving up more of our waking hours to read, to study the facts, examine the truths and weigh the possibilities, and to accept the potential loss of personal gain in exchange for the advancement of the whole. The history of the world clearly reveals that much of the success of virtually every innovation or reforming enterprise lies in the ability of its leaders and innovators to apply rigor, take risks, and make sacrifices for others. They were the hallmark behaviors of America's founding fathers and the put-upon colonialists, have been at the core of every successful entrepreneurial enterprise and are the headlines that sit on top of every corporate turnaround story since the beginning of time. Achieving effective, successful solutions to our most vexing and endemic problems require that many more of us be willing to step forward and sacrifice our time and dedicate our attention to really understanding every aspect of the systems we need to improve, the challenges we and others are attempting to solve, and the human code that is both the underlying problem and the likely solution. Central to the task is a willingness to question all the pre-existing conditions and legacy assumptions within us and our institutions and develop plausible pathways of experimentation, trial and error that will enable us to iterate towards our definition of a better future.

The original human code is first and foremost a language of feeling and intuition, a language built of behaviors and biases that grasp at logic as self-serving validation and that largely operates in monologue form, with nominal consideration of what the facts might have to say. As Daniel Kahneman depicted, System 1 too often fools System 2. For humankind to prevail over the foes and forces we face, more of us must accept that to effectively translate that language, to benefit from its meaning, we must present the opposite, a dialect of rigor, objectivity, and truth. System 2 must convince System 1 more often. It is the dialect of the intellectually engaged and humanly introspective, conveyed through a structured process that serves as a sacrificial, unselfish quest to find the answers that truly reflect who and how we are, answers that work for humankind, despite ourselves.

In addition to closing the gap between technology's advance and our own, our structured process must carefully catalog and address the myriad of unintended and negative consequences that technology unleashed as wrought on the world. Such consequences are magnifiers of the dangers of the gap and the importance of closing it quickly.

Human behavior is the root of all actions, decisions, and consequences.

II

The Unintended Consequences

CHAPTER 4

The Emerging
Specters of Threat

"There can be no Plan B because there is no planet B."

UN Secretary-General Ban Ki-moon

Technology has changed many things, except for one big thing: our reluctance to change our behaviors. Our fear of strangers, our instincts towards gaining advantage for self and our tribe, and our baked in Maslovian needs for survival and control, are joined by one other unpleasant human truth, our thirst for battle. We have forever sought foes and been seduced by the thrill of conquest. We are and have always been wired for war. The Neanderthals battled other Neanderthals, the Ancient Greeks took on Persia one too many times, and the Christians endured multiple Crusades to vanquish the infidels. As I write this, there are at least 28 global conflicts happening, and that does not include the unregistered or unacknowledged internecine wars of persecution and genocide. The evolution of modern civilization has not brought with it a human willingness to let go of war. According to an article in *The New York Times* in 2003, "Of the past 3,400 years, humans have been entirely at peace for 268 of them, or just 8 percent of recorded history." Our continuous warring ways are partly reflective of our compulsive quest for more and a belief that human progress is measured by how much we have. The more goats, the more power. There is also within it a form of individual manifest destiny. Many if not most of us are programmed to compete, to win. Our survival instincts easily slip into a desire to conquer or watch others conquer. (It's why the global professional sports industry is a $500 billion business, attracting billions of followers and fans. Perhaps

DOI: 10.1201/9781003089902-7

not coincidentally, the global gambling industry, the industry of picking winners, is the same size, plus or minus a few billion dollars.) The bad news and good news are that humans tinkering with technology have invented something called nuclear bombs, devastating devices that while existential threats in themselves, are arguably effective deterrents to the prospects of a third world war. Even the most ardent of warring nations understand the irrevocable and horrifying consequences of hitting the button.

A DESIRE FOR GREATER STANDING

Opposing sports teams as enemies aside, our warring foes have evolved from being other tribes, to our own countrymen, to other countries, and alliances of other countries. Across the history of humankind, wars have been waged based on religious differences, the loss of freedoms, and a myriad of seemingly disparate root causes. In his book *Why Nations Fight*, Richard Ned Lebow consolidates that view, citing five common motives: "security, material advantage, standing, revenge and domestic politics." In his analysis of every war since 1648 he determined that the presumed primary motives, security and material advantage were only responsible for a handful of conflicts and that standing and revenge were actually the biggest instigators. Consider the current war in the Ukraine, provoked by Vladamir Putin, Russia's dictatorial and megalomaniacal president. Even with his initial nationalistic claims for greater border security, in truth, the incursion is more likely a matter of standing and revenge.

Lebow goes on to characterize these two psychological motives as "spirit"-based, suggesting that the nature of war is a direct function of the nature of man, that it may be less about what we want and more about what sense of place we seek and who has pissed us off. We may want more goats, but we really want more respect and greater status. We are back to Maslow's Hierarchy of Need and our incessant, never flagging quest for validation of self. Lebow writes,

> In modern times the spirit (thumos) has largely been ignored by philosophy and social science. I contend it is omnipresent. It gives rise to the universal drive for self-esteem, which finds expression in the quest for honor or standing. By excelling at activities valued by our peer group or society, we win the approbation of those who matter and feel good about ourselves.

In his 2021 book *The Status Game*, author Will Storr paints the picture even more vividly:

> No matter where you might travel, from the premodern societies of Papua New Guinea to the skyscraper forests of Tokyo and Manhattan, you'll find it: humans forming groups and playing for status…within these groups, we strive for individual status—acclaim from our co-players. But our groups also compete with rival groups in status contests: political coalition battles political coalition; corporation battles corporation; football team battles football team. When your games win status, we do too. When they lose, so do we. These games form our identity. We become the games we play.

Whatever the game or context, achieving status and standing in Maslovian terms is a form of self-actualization or at least elevated self-esteem. War, like technology itself, is a blunt and brutal instrument for achieving those outcomes, and always brings with it a slew of unintended and dark consequences.

A DIFFERENT ENEMY, A DIFFERENT THREAT

For centuries the traditional model of war was focused on enemies who looked remarkably like us. Since the first club was swung by a human at a human, our competitors have had familiar characteristics. The other side was visible, describable, and discrete, to the point of wearing red coats to help us shoot the right people. The wars we have waged, the enemies we have pursued or defended against have always been bound by time, space, resources, and rules. These enemies were physical entities that had measurable capacities and identifiable weaknesses who in the modern age often agreed to fight within generally accepted confines of morality and fairness and the implicit rules of international law.

Throughout history there has also always been a beginning, middle, and end to wars, and until the last 50 years or so, the outcome has mostly been binary, win or lose, with an occasional and entirely unacceptable draw. Think Vietnam, Iraq, and Afghanistan. And now, probably, the Ukraine and what's happening in the Middle East. Thanks in part to technology, there is now the ability to wage a "non-total war," to arrive on the doorsteps of the foe with an agenda not to vanquish but to democratize or quell. In the traditional win or lose version of war the greatest determinants of victory have been simple: the size of each side's armed forces, their resources, training, weaponry, and technological advantages, the cunning applied to the strategies of battle and overall conquest, and the degree of intensity the opposing forces brought to the battlefield. The longstanding nature of war reflected the longstanding nature of man. The supporting construct was guided by age-old formulas and a largely reactive, remedial stance, a game of chess where the enemy's next moves could be contemplated and calculated. Occasional plagues aside, the greatest threat to our existence has always been a foe that looked like us and played chess like us, and sometimes just played it better. And perhaps our own human nature that couldn't resist creating or entering the fray to begin with.

Once again, technology has changed the game. The rapid, unchecked advance of the technology train has brought with it some unintended passengers, a new kind of mostly invisible foes that increasingly represent existential threats and carry the potential to accelerate the decline of civilization and the destruction of our planet. This is an entirely unpleasant thought I know, but factually supported. Today finds our species flat-footed, facing a slate of new enemies, bearing characteristics that are effectively the opposite of what we have come to know. These emerging specters of threat are unseen, amorphous, and unwilling to abide by conventions of war. And they are wide-ranging, from pandemics and cyber-terrorism to climate change and antibiotic drug resistance. Regardless of form, they are invasive threats unbounded by time, space, resources, or law. The duration of battle with each is likely to be infinite, requiring a re-definition of winning, because in most cases these foes cannot be entirely vanquished, and in many cases our new enemies are intrinsically connected, resulting in an informal and yet formidable alliance against us. Again, unpleasant but true. (Of note, in his compelling book *Precipice*, author Toby Ord separates the existential threats we face into two camps: NATURAL RISKS, e.g., super volcanic disruptions and stellar explosions versus ANTHROPOGENIC RISKS, those that are human-caused or magnified. My focus is on the latter because we can do something about them.)

A DIFFERENT BATTLEFIELD, A DIFFERENT OUTCOME

As difficult as this is to consider, our expectation, read hope, of an absolute outcome, of winning, against our unseen enemies is no longer reasonable. The chessboard should be put away, to be replaced with an entirely new model of enemy engagement. It is a model that is borderless, less about finite capacity and resource limits and more about infinite collaboration and global coordination. The definition of winning, when winning is not an attainable end, requires a massive shift in orientation from time-framed elimination and fighting to a conclusion, to perpetual prevention and remediation. The new battlefield strategies and actions must be focused on disabling and diminishing future attacks while preparing for the inevitable next round. Combatting the myriad of invisible, borderless threats will require unprecedented levels of international agreement on multi-lateral protocols and standards that challenge the ruling freedoms of sovereign states and nations. Such collaborations might call for a global governing system that presents codified approaches, requires shared sacrifices at the individual and collective levels, and demands both compromises and bold new innovations on every front. Paradoxically, as technology has played a lead part in creating or magnifying these threats it also must play a major role in mitigating the damage those threats can do.

The battle fought against COVID-19 was a perfect petri dish for examining how technology has helped create and exacerbate a global threat, how humankind struggles to establish and adopt the new rules of the game, and how technology could help, but only if humans were willing to change their ways. The coronavirus was not our first pandemic, nor will it be our last. But the main difference this time was the speed at which the virus moved around the world, in part due to the explosive growth of air travel over the last 20 years and the technology that enabled it. (Whether the virus escaped a research lab remains subject to debate.) According to a study by the Rand Corporation, the worldwide spread of COVID-19 due to air travel was happening weeks before the World Health Organization realized it had a pandemic on its hands. They reported,

> In particular, we estimate that the worldwide exports of COVID-19 cases began increasing at an accelerating rate on February 19, 2020, exactly three weeks before the WHO declaration. By the end of February, about two weeks before the WHO declaration, more than five cases of COVID-19 per day—or nearly 40 per week—were already being exported around the globe via air travel.

As air travel rapidly spread the disease, another technological "advance," social media, spread misinformation. Across the globe, rumors, magic elixirs, and outright lies were being proffered and perpetuated by government leaders, scientists, and everyday citizens. The result was massive confusion, a further erosion of societal trust, and growing doubt that the pandemic was really worth worrying about. As the misinformation swirled so too did most countries' governments and their attempts to put together a pandemic response battle plan. Even with China revealing an initially effective method of shutting down the virus by shutting down their society for 76 days, the rest of the world, and the developed world in particular, was reluctant to take such draconian steps. And the question must be asked, why? It's likely a consequence of many factors, ranging from the structural to the psychological. At the core is the Chinese system of government and a general disdain felt by democratic nations for its authoritarian rule and the way it subjects its citizenry to control and coercion. While China was held to blame for the virus, its effective short-term prescription was almost universally rejected as too harsh and requiring too much sacrifice of both the country's economic health and the quality of its peoples' lives. In commenting on China's response, *The New York Times* science and health reporter Donald G. McNeil Jr. pushed back on that stance, saying,

> But there are a lot of brutal things that the government in Beijing does. In this case, it was not brutal to its own citizens. It saved probably 10 million lives. That's how many I estimated would have died in China if this had just gone unchecked.

The Chinese government demanded sacrifice from its citizens to save their lives, something the rest of the world was loath to do. It's a scenario that should be contrasted with how many nations behave during wartime, beseeching their citizens to step forward to give up freedoms and comforts to help vanquish the enemy for the benefit of all. During World War II, the American and British rationing of sugar, gas, steel, and a range of other commodities was applied without hesitation, supported by the consensus recognition that beating back the Axis powers would require every citizen to contribute. During the pandemic governments were hesitant to ask for sacrifice because we could not objectify this enemy, the potential loss of life was not universal, and perhaps because in many democracies, individual freedoms have become sacrosanct and even weaponized. Individualism now takes precedence over the good of the whole.

THE VOID OF GOVERNANCE

The varying structures of governments have clearly had a major impact on how variable the response to the pandemic was. Federated systems like that of the United States revealed a profound weakness in the country's ability to negotiate and dictate the appropriate protocols for mitigation and response to an invisible, borderless enemy like COVID-19. The dance between the span of control and decision-making capacities of the Federal government versus that of the 50 states resulted in a confusing merry-go-round of half-policies and half-measures that were deployed inconsistently and largely ineffectively. The result: the world's number one superpower became the pandemic's biggest hotspot. As of 2023 the world topped 770 million cases, the United States passed 1.1 million COVID-related deaths, the most deaths of any country, and a new Omicron variant was just revving up. Even though effective vaccines have been available for 3 years, the death toll and infection rates continue. Individual freedom, a lack of collective responsibility and misaligned forms of global governance didn't work so well together against an invisible, borderless, and pernicious enemy.

The biggest contributor to the global reluctance to follow China's path and really "lock down" to stop the virus spreading, however, may have been the simple matter of belief. We humans tend not to believe what we cannot see or do not personally feel. It is hard for us to accept a collective threat when the threat is invisible and not actually at our door. And even when we saw images of cold storage units serving as temporary morgues for the growing COVID-19 fatalities, we didn't see dead soldiers or casualties of war. Instead, we saw unlucky outliers. As long as we were able to live our lives, the problem was not really a problem, and no sacrifice was required. Alongside such persistent disbelief was and is a tendency towards remediation versus prevention, and a willingness to pay only when we absolutely must. Ezra Klein, then the editor-at-large for Vox once wrote about this persistent, problematic human behavior vis-a-vis climate change, particularly in America:

> The pain of doing something serious about the problem is upfront. But the worst effects of global warming won't be visible, even in America, for a long time to come. The true crisis is abstract while the sacrifice required to prevent it is tangible. The American political system (or citizenry) is not good at trading sacrifice now to prevent crises later.

As we struggled to find and embrace the new rules of the COVID-19 game, a game made harder by technology, we also saw the capacity of technology to help us win it, or at least try to get the pandemic in check so we can prepare for the next one. The greatest evidence was the speed at which viable COVID-19 vaccines were developed. Prior to this pandemic the generally accepted view was that an effective vaccine development process took 10 to 15 years to develop, and even then, there was no guarantee of success. Decades of research and trials later, there still is no vaccine for malaria, a disease that impacts hundreds of millions of people every year and kills hundreds of thousands, mostly young children. The COVID-19 vaccine development process was entirely different. Eleven months after the virus was first detected two vaccines were made available to parts of the world, both with 95% effective rates and both developed and clinically trial-tested in 250 days. A remarkable achievement, it is a direct consequence of not just scientific advances but of the public health effort across multiple countries. Public and private sectors collaborated and coordinated in a shared search for a vaccine to combat this universal foe. We can fight these borderless threats together. We just need to do more of it.

(Of note, in contrast to China's initial COVID-19 response success, it continues to struggle with the outbreaks, due in part to less effective vaccines and the societal and economic consequences derived from its Zero COVID policies.)

In 2021 Researcher B.S. Manjunath wrote a piece on the pandemic for the *Economic Times*,

> What we actually need is preparedness. Indeed, the technology has advanced more and will continue to advance exponentially, but the human institutions and societies need to accelerate in adapting to it and continue investing in building the technology systems for the preparedness. After the COVID-19 outbreak, it is evident that, from AI to robotics, the technology innovations are helping to manage the epidemic and better equip to fight future public health emergency in a timely, systematic, and calm manner.

The subtle inference here is that while technology is critical it must be accompanied by a human commitment to systematic and steady collaboration among nations. Collaboration is the only viable means to take on the growing list of invisible enemies, including a relatively new threat, known as cyber-terrorism.

TWO NEW KINDS OF PANDEMIC

Cyber enemies are not dissimilar from biological viruses. They are invisible, ignore geographic borders, and much of humankind does not believe that the threat is really a threat. The machinery of the entire global economy would come to an instant economic depression-inducing halt if these hidden enemies were able to disable the Internet and all the systems that now rely on it. Cyber-terrorists have the capacity to severely disrupt countries, institutions, companies, and the lives of individual citizens through any number of hacking means, motivated by ideological beliefs or just the perverse desire to create havoc around the world. And the world has never experienced an enemy like this. In his presentation to America's Congressional House Homeland Security Committee in 2019, then Director of the Federal Bureau of Investigation Christopher Wray said this:

> Virtually every national security threat and crime problem the FBI faces is cyber-based or facilitated. We face threats from state-sponsored hackers, hackers for hire, organized cyber syndicates, and terrorists. On a daily basis, these actors seek to steal our state secrets, our trade secrets, our technology, and the most intimate data about our citizens—things of incredible value to all of us and of great importance to the conduct of our government business and our national security. They seek to hold our critical infrastructure at risk, to harm our economy and to constrain our free speech.

As threatening as this enemy is, like COVID-19, there remains a remarkable combination of disbelief and disregard among millions of people in terms of what danger really exists and whether they should do anything to protect themselves from it. In a recent Pew Research study, most American adults surveyed acknowledged that their personal data was likely less secure than it was 5 years ago, and yet, they also revealed that they are increasingly less diligent about understanding privacy policies, deploying two-step security verifications, and applying best practice thinking to password formation to protect their identities and personal assets. The theory holds. We do not fear what we cannot see or feel. While COVID-19 appears to be at least subdued, cyber enemies cannot be. And because the World Wide Web is exactly that, global and connected, it means that our ability to push back the enemy's unrelenting assault and protect our assets, big and small, should be addressed globally, in a connected, systematic fashion and in perpetuity. But individuals need to also step forward and replace disbelief and disregard with fact-based commitment to doing their part to mitigate the risk to them, to others and for future generations. And even with that collective effort, our grandchildren's grandchildren will inevitably be fighting cyber wars. They will be fighting a foe that was created or at least enabled by technology and their weapons will need to be technology, global collaboration and their own individual responsibility.

It's a dynamic that also applies to drug resistance, another pandemic-like enemy that will eventually kill millions unless we can coordinate our response. As the World Health Organization has declared,

> Where antibiotics can be bought for human or animal use without a prescription, the emergence and spread of resistance is made worse. Similarly, in countries without standard treatment guidelines, antibiotics are often over-prescribed by health workers and veterinarians and over-used by the public. Without urgent action, we are heading for a post-antibiotic era, in which common infections and minor injuries can once again kill.

Fundamentally we have over-prescribed, over-distributed, and over-consumed antibiotics because we could and because technology made it easy and convenient. And now we are facing the consequences and hoping that new technologies and changes in human behavior will somehow save the day.

POPULATION AS THREAT

There is another encroaching and invisible foe born from the combination of increasing life spans and declining birth rates, both directly and indirectly a function of technology. Technology has undoubtedly extended lives through advances in medicine and increased access to food and water and basic health care. According to data from the United Nations Population Division, global life expectancy has increased from 47 years in 1950 (both genders) to 73.2 years in 2020, a remarkable 55% improvement. But technology has done something else. It has enabled millions of women in the developed world (and increasingly developing world) to access improved forms of contraception like IUDs and patches, to obtain higher and higher levels of education, and to participate actively in the workforce. The combination of those factors and a subtle but real societal shift away from

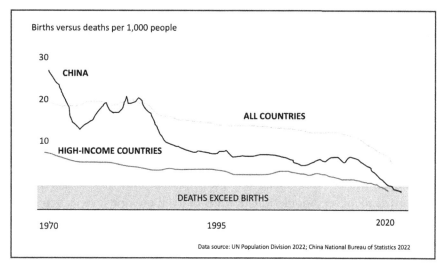

Births versus deaths per 1,000 people

30

CHINA

20

ALL COUNTRIES

10

HIGH-INCOME COUNTRIES

DEATHS EXCEED BIRTHS

1970   1995   2020

Data source: UN Population Division 2022; China National Bureau of Statistics 2022

FIGURE 4.1   Global population trends.

child rearing as an essential adult function, has resulted in declining birth rates across much of the developed world and in China, as reflected in this stunning graph derived a *New York Times* article and population data from the United Nations and China (see Figure 4.1).

As reported by James Gallagher, the Health and Science correspondent for the BBC News, most developed countries will see population declines of as much as 50% by the end of this century. He writes,

> Japan's population is projected to fall from a peak of 128 million in 2017 to less than 53 million by the end of the century. Italy is expected to see an equally dramatic population crash from 61 million to 28 million over the same timeframe. They are two of 23 countries—which also include Spain, Portugal, Thailand and South Korea—expected to see their population more than halve.

America is experiencing the same, albeit not so dramatic population trend. A U.S. government census report states, "…by 2034, we project that older adults will outnumber children for the first time in U.S. history." According to United Nations and Pew Research studies total worldwide population growth is projected to hit 9.7 billion people by 2050 and top out at 11 billion by the end of the century, reflecting a disproportionate growth in developing countries while the developed world shrinks and ages. The economic and geo-political consequences of these shifts are almost infinitely complex.

At first glance, declining populations in some countries, given the other invisible threats associated with environmental degradation and the ongoing challenges to feed an ever-hungrier planet, would be viewed as a net positive. But as Gallagher later pondered, "Who pays tax in a massively aged world? Who pays for healthcare for the elderly? Who looks after the elderly? Will people still be able to retire from work?" Ironically, while we are witnessing an alarming rise in nationalistic fervor across the developed world, study after study show that the best way for developed countries to offset population decline is to allow and even encourage greater levels of immigration, because immigrant families tend to have more children. In the conclusion of Gallagher's piece, Professor Ibrahim Abubakar of the University College London is quoted "If these predictions are even half accurate, migration will become a necessity for all nations and not an option… . The distribution of working-age populations will be crucial to whether humanity prospers or withers." The unintended negative consequences of declining birth rates and increasing longevity, when combined, are stark and potentially devastating. And yet our ability and willingness to grasp that fact and to counteract it with the requisite policies and behaviors *now* to offset the likely future devastation is limited. Our concern and activism is diminished by the threat's invisibility, a decades-long time horizon and again, the recurring behavior that we do not fear and will not act on what we cannot see or feel. Most fundamentally, we appear incapable of proactive prevention or even mitigation of the growing threats. Knowing what we now know of the future population distribution curve should be sufficient motivation for an intensive overhaul of our core systems, beginning with health care and including immigration policy. A world within which only 10% of the population works, or can work, demands a complete rethink of the role of government, politics, and even how economics must be re-fashioned to deliver a sustainable future for all.

TICKING NUCLEAR TIME BOMBS

The rise and fall of collective concern regarding the potential of nuclear threats, from bombs to energy plant waste storage, reflects our all too human capacity to forget or avoid the things we really don't want to deal with. Just after World War II, nuclear weapons were at the top of our existential threat list. The devastation in Hiroshima and Nagasaki was terrible proof that we had created a potential enemy that could kill every one of us. By 1965 an estimated 200,000 backyard bunkers had been built across America. The number is a rough estimate because given the level of urgency at the time, permits were not required. Fast forward to 2022 and most people would not put nuclear weapons on the most threatening list, even though there are enough nuclear armaments scattered around the world to annihilate all its occupants.

In *The Precipice*, author Toby Ord digs deep into the nuclear threat, underscoring that the existential threat issue is less the direct impact of a nuclear hit, and more the derivative consequences, what has been dubbed a "nuclear winter." He writes,

> It wasn't until the early 1980s—almost forty years into the atomic era—that we discovered what is now believed to be the most serious consequence of nuclear war. Firestorms in burning cities could create great columns of smoke, lofting black soot all the way into the stratosphere. At that height it cannot be rained out, so a dark shroud of soot would spread around the world. This would block sunlight: chilling, darkening and drying out the world. The world's crops would fail, and billions could face starvation in a nuclear winter.

He's describing a picture surprisingly like what might happen when a super volcano erupts. But in the nuclear case, humans would have made it happen. Ord goes on to point out that as alarming as that prediction may be, we have made progress reducing the nuclear threat; that the risk has significantly decreased thanks to a globally negotiated reduction in nuclear arms and their impact capacities. His qualifier is that even with those reductions, the increase in geopolitical conflicts, e.g., Russia's invasion of the Ukraine and by extension Europe and the United States, could result in a renewed nuclear arms race and with that another global enemy to contend with.

AI: FRIEND OR FOE?

Technology is without question the creator of the nuclear threat, the enabler of the other invisible threats, and the threat itself in the case of rapidly emerging areas like Artificial Intelligence (AI) and robotics. The generally shared intellectual fear is that humankind's survival or at least position at the top of the food chain is vulnerable to the rise of the machine and a future computer's capacity to out-think and out-perform us, whether in a job function or as a ruling entity. There is a legitimate concern among many in the science and business communities around the world not just at the potential for massive job losses from AI-induced automation and robotics, but rather that computers will one day beat us at the game; not just at chess, but at the game of running the world. In one episode of his provocative streaming series, *This Giant Beast That Is the Global Economy*, the actor and former White House Associate Director of Public Engagement Kal Penn takes an engaging look at the realities of AI and the possibility that once fully unleashed the consequences could be devastating to humankind. Across a range of Penn's interviews, it becomes all too clear that even the professionals at the forefront of stewarding our understanding and application of AI and its derivative Deep Learning and Machine Learning capacities do not and perhaps cannot predict how these game-changing technologies will impact the world. In a conversation with Andrew McAfee, the Co-Director of the IDE and a Principal Research Scientist at the MIT Sloan School of Management, Penn explores the likelihood of AI creating another industrial revolution. McAfee points out that as the first revolution was fueled by the invention of steam, electrical, and combustion engines to address the limits of muscles, an AI-fuel revolution would address the limits of the human brain. And he goes on to recognize that while revolutions bring wholesale and often positive changes to the way things and humans work, they also tend to carry significant human costs. The problem is that no one knows what those costs might be and therefore whether the gains are worth the losses. The dilemma is perfectly captured in a short clip quoting the theoretical physicist, cosmologist, and author Stephen Hawking, likely taken from a speech he gave at a conference in 2017:

> Success in creating effective AI, could be the biggest event in the history of our civilization. Or the worst. We just don't know. So, we cannot know if we will be infinitely helped by AI, or ignored by it and side-lined, or conceivably destroyed by it.

Effectively we do not know if AI and its myriad spawn, including ChatGPT, are important friends or the ultimate enemy. Most of us cannot see AI, most of us do not understand it, and we cannot imagine what this potential foe might do to our lives and our livelihoods. And yet with all those unknowns, and all the risks they carry, science, technology, and business continue to sally forth pushing the AI agenda, albeit with an emerging effort to establish "ethical AI" principles as a means of keeping this new and powerful partner in check. Over the last decade, a range of countries, companies, and alliances, from the European Union and the OECD to Google and Microsoft, have banded together to create guidelines that attempt to mitigate the risks and the unintended ethical and functional consequences of an increasingly AI algorithm-driven world.

On October 30, 2023 the Biden White House issued an executive order to attempt to corral the potential dark side of Artificial Intelligence, declared as "the most sweeping actions ever taken to protect Americans from the potential risks of AI systems." The order called for six major actions:

- *"Require that developers of the most powerful AI systems share their safety test results and other critical information with the U.S. government. In accordance with the Defense Production Act, the Order will require that companies developing any foundation model that poses a serious risk to national security, national economic security, or national public health and safety must notify the federal government when training the model, and must share the results of all red-team safety tests. These measures will ensure AI systems are safe, secure, and trustworthy before companies make them public.*
- **Develop standards, tools, and tests to help ensure that AI systems are safe, secure, and trustworthy.** *The National Institute of Standards and Technology will set the rigorous standards for extensive red-team testing to ensure safety before public release. The Department of Homeland Security will apply those standards to critical infrastructure sectors and establish the AI Safety and Security Board. The Departments of Energy and Homeland Security will also address AI systems' threats to critical infrastructure, as well as chemical, biological, radiological, nuclear, and cybersecurity risks. Together, these are the most significant actions ever taken by any government to advance the field of AI safety.*
- **Protect against the risks of using AI to engineer dangerous biological materials** *by developing strong new standards for biological synthesis screening. Agencies that fund life-science projects will establish these standards as a condition of federal funding, creating powerful incentives to ensure appropriate screening and manage risks potentially made worse by AI.*

- **Protect Americans from AI-enabled fraud and deception by establishing standards and best practices for detecting AI-generated content and authenticating official content.** *The Department of Commerce will develop guidance for content authentication and watermarking to clearly label AI-generated content. Federal agencies will use these tools to make it easy for Americans to know that the communications they receive from their government are authentic—and set an example for the private sector and governments around the world.*

- **Establish an advanced cybersecurity program to develop AI tools to find and fix vulnerabilities in critical software,** *building on the Biden-Harris Administration's ongoing AI Cyber Challenge. Together, these efforts will harness AI's potentially game-changing cyber capabilities to make software and networks more secure.*

- **Order the development of a National Security Memorandum that directs further actions on AI and security,** *to be developed by the National Security Council and White House Chief of Staff. This document will ensure that the United States military and intelligence community use AI safely, ethically, and effectively in their missions, and will direct actions to counter adversaries' military use of AI."*

At first blush Biden's executive order is a major step forward in our attempt to steward this world-changing technology. But when examined through a more rigorous lens, it lacks sufficient connection to or acknowledgement of the human part of the equation. The references to safety don't go far enough to acknowledge how this technology, like many technologies, can and likely will prey on our psychological vulnerabilities. How AI could seduce and manipulate us to do the wrong things, even bad things, without us knowing it. The order and its component directives need far more human grounding to be effective. It also needs two other things it does not have: enforceability in America and global application. Remember, AI knows no borders. So we have a ways to go.

In truth, the challenge within the task of creating any effective guidelines and guardrails for responsible AI is once again, an issue of complexity. As captured in her provocative 2019 article in *Fast Company*, anthropologist and senior researcher Sally Applin declared,

All of this means that the "ethics" that are informing digital technology are essentially biased, and that many of the proposals for ethics in AI—developed as they are by existing computer scientists, engineers, politicians, and other powerful entities—are flawed, and neglect much of the world's many cultures and ways of being. For instance, a search of the OECD AI ethics guidelines document reveals no mention of the word "culture," but many references to "human." Therein lies one of the problems with standards, and with the bias of the committees who are creating them: an assumption of what being "human" means, and the assumption that the meaning is the same for every human.

Applin exposes a fundamental and perhaps uncomfortable truth about our ability to deal with complexity to both steer technology and countermand its many related threats. To do so, to win or to simply not lose, we must become much better at understanding us and defining our desired attributes and acceptable behaviors. But even if it were possible to establish such absolute measures of right and wrong and implement a one-planet, truly global systems of standards and enforcement, it would likely still be insufficient. AI is increasingly ubiquitous and available, its tools so widely distributed and its experiments so low cost that any rogue actor or nation anywhere could create a malicious AI force by design or by accident that would unleash a set of problematic, potentially existential consequences. Like a pandemic or nuclear fallout, AI does not care about borders. And like a pandemic, even a temporary cure is complicated to create. Applin concludes that ethical AI is a multi-faceted, human problem to solve:

> Artificial intelligence must be developed with an understanding of who humans are collectively and in groups (anthropology and sociology), as well as who we are individually (psychology), and how our individual brains work (cognitive science), in tandem with current thinking on global cultural ethics and corresponding philosophies and laws. What it means to be human can vary depending upon not just who we are and where we are, but also when we are, and how we see ourselves at any given time. When crafting ethical guidelines for AI, we must consider "ethics" in all forms, particularly accounting for the cultural constructs that differ between regions and groups of people—as well as time and space.

As technology has helped foment the many specters of threat, it must also play a role in mitigating their impact on humanity. It is possible that the greatest application of Artificial Intelligence may one day be its capacity to address the seemingly infinite complexities of the ethics question, resulting in an ability to police itself, not just in the near future but forever.

ONE PLANET

This brings us to what should be considered the threat of threats, the poster child for this burgeoning list of invisible, largely unvanquishable, and mostly unbelievable enemies: the growing ecological imbalances throughout our planet's ecosystem. Contrary to most of our belief systems and actions, we are part of one natural order ecosystem. The balanced health of that combined ecosystem determines the future health and viability of all humankind. As noted in a 2019 United Nations report, "All species, including humans, depend for their survival on the delicate balance of life in nature." The growing imbalance within our planetary environment has several inter-connected facets, sub-threats of a kind, that are beginning to wreak havoc on our capacity not only to thrive but to survive. And unlike prior ecosystem-changing, extinction-inducing periods, including the Cretaceous Period 66 million years ago, when most of all species and dinosaurs disappeared, this period of planetary change is largely man- and technology-made and it carries very real consequences. In the presentation of a landmark report in May of 2019, the head of the Intergovernmental Science-Policy Platform on Biodiversity and Ecosystem Services (IPBES), Sir Robert Watson declared,

> The overwhelming evidence of the IPBES Global Assessment, from a wide range of different fields of knowledge, presents an ominous picture... .The health of ecosystems on which we and all other species depend is deteriorating more rapidly than ever. We are eroding the very foundations of our economies, livelihoods, food security, health and quality of life worldwide.

It's a view corroborated in the World Economic Forum's *The Global Risk Reports 2020* and its identification and mapping of the top 30 threats to humankind against two critical variables, the degree of impact and the likelihood of the threat happening. Of the 30 threats, the first six of the top seven are environmental and are at the very least mutually causal, and largely all caused by the mother of all our problems, climate change and our inability to unilaterally respond.

(1) Climate action failure

(2) Extreme weather

(3) Biodiversity loss

(4) Natural disasters

(5) Human-made environmental disasters

(6) Water crises

(7) Cyber-attacks

The degradation of the health of our planet, the declines in its functioning as an efficient and self-sustaining system, and the accelerating extinction of species are all the spawn of decades-old abuses made more extreme by the double whammy of technology and the self-interest and greed of humans. Largely on the back of technology enabling the excess of access, unchecked consumption is leaving us bloated and the planet itself depleted. As active participants in the ravishing of Earth's finite resources, humans have long displayed an ignorant belief that there will always be more or enough, alongside a blatant disregard for the downstream consequences of our actions to the environment and its non-human inhabitants. In a recent speech at Columbia University, U.N. Secretary-General Antonio Guterres declared, "The state of the planet is broken. Humanity is waging war on nature. This is suicidal." The techno-utopians would decidedly disagree. Technology will solve the problem they say. And it might or might not.

The hubris, naivete, selfishness, and denial of the degree of the environmental threats apply to the other myriad of invisible threats and the linkage between them. *The Economist*, citing another IPBES panel presentation this past year, described the correlation between the threat of future pandemics and the accelerating loss of biodiversity:

An expert panel warned this week that not only are pandemics becoming more frequent, but unless they are handled differently, they will spread faster, kill more people and do more damage to the global economy. The destruction of natural habitats as well as the trade and consumption of wildlife are to blame. Both lead to closer contact between humans and animals, making cross-species infection more likely. The panel estimates that birds and mammals are host to 1.7m yet-to-be discovered viruses, of which 540,000-850,000 might affect humans. It prescribes a shift from reactive pandemic responses to approaches that reduce the risk of future outbreaks, such as habitat conservation. Serendipitously, those approaches could help in the fights against climate change and biodiversity loss, too.

The links between our treatment of the environment and the severe consequences to humankind are undeniable. Consider what we have done with plastics, what plastics have done to Earth, and what that is doing to us. According to a 2022 research study funded by the Minderoo Foundation, plastic-related global health issues cost over $100 billion per year.

If we look ahead objectively, and assume insufficient corrective actions, we will likely see a world that for all its technology-enabled efforts to prolong human life has ultimately actually done the opposite. There is ample predictive scientific research that paints our future planet as a place where entire swaths of countries are largely submerged while others are destroyed by a lack of water; where the growing frequency and intensity of hurricanes and typhoons force millions of people to abandon their homes and livelihoods; where the pollution of our air and water has passed a point of no return killing thousands of species and threatening human health; leaving a humankind which cannot feed itself because it has fed too much off the planet that gave it life.

It and we are all connected. What we do to the planet we ultimately do to ourselves. The Global Risks Report 2020 calls out a stunning cause and effect between the decline in insect populations due to climate change and the rise of human health epidemics like obesity:

> Insects are also the world's top pollinators: 75% of the 115 top food crops rely on animal pollination, including nutrient-rich foods like fruit, vegetables, nuts and seeds, as well as cash crops such as coffee and cocoa. Dwindling insect populations will force farmers to seek alternative means of pollination or shift to staple crops that do not rely on pollinators. However, these crops—such as rice, corn, wheat, soybeans and potatoes—are often energy-dense, nutrient-poor and already over-consumed globally, contributing to an epidemic of obesity and diet-related disease.

We need no more evidence to understand the linkage, the problems, or their solutions. But much of humankind still refuses to accept that the threat of environmental degradation and the growing imbalance in our planet's ecosystem, is an enemy that we should take seriously now, an unrelenting force that demands both individual and collective sacrifice and new approaches to playing this life-or-death game. Once again, our reluctance to act is largely because the threat is invisible, its current impact intermittent and often distant, and because we are biologically programmed to reject the idea that this enemy might conquer us when we know we can't really conquer it.

Charles Homans, the political editor at *The New York Times*, perfectly captured our persistent DNA-wired threat denial capacity in his article about the late 2019 Australian fires titled "A Disaster Video That Finally Tells the Truth About Climate Change." He depicted how most humans process the horror and devastation that the climate change-induced conflagration wrought through two different parts of the brain, effectively in sequence. The frontal lobe sees the video footage and maybe some of the stunning loss of life statistics and determines, "That is really bad.", "We need to take steps to address a growing problem." Then the amygdala kicks in, the tiny little gland in the center of our brains that is best known as the instigator of our "fight or flight" orientation. The amygdala quickly overrules the frontal lobe, convincing us that it's not our problem, that we've seen similar things happen and people survived, and that since we're individually not at risk, why worry about it. Homans writes.

> For most of the time we have understood it as an existential threat, climate change has offered an extreme version of the phenomenon that economists and psychologists call risk discounting. This is the degree to which our sense that a risk is distant — in space or, especially, in time — affects, consciously or otherwise, our assumptions about its severity. The fear has always been that by the time palpable evidence of the threat arrives, and we are sufficiently alarmed to do something about it, it will be too late.

The fear Homan calls out is not yet a true fear for most of us. It's really not. He concludes by sharing, "We instinctually envision climate change as a series of disaster stories with recognizable beginnings, middles and endings, rather than as a single disaster story that outlives us, maybe all of us." Like many of the other specters of threat, the seemingly intermittent and often invisible nature of the enemy called climate change has no end. The only option open is mitigation of risk, in which we concertedly shift from being abusers of our only ecosystem to its active and activist stewards committed to its sustainability. It is that shift and that shift alone which will ensure the sustainability of the human race. Whether the specter of threat is climate change, future pandemics, negative birth rates, Artificial Intelligence, or nuclear war, we must learn to play the game differently in every facet of human society and the systems that enable it. Exploitation and the need to win must be replaced with cultivating, managing, and preventing. The concern for borders must be replaced with an unwavering commitment to global collaboration. And most importantly technology should only be deployed with direct and primary consideration of the human factors involved and at stake. It's a completely different game that requires intense focus on the linkage between technology, our behaviors and our actions, and the myriad of unintended consequences it and we perpetrate. And how our decisions have a domino effect capacity hidden in the future that may be beyond even our best science to be able to calculate. Finally, we all need to accept that as the players of this new game our ignorance of the rules, of the human truths and our unwillingness to collectively and individually confront these threats is the greatest invisible threat of all.

> The enemy is what we cannot see; what we cannot understand; what we will not accept.

CHAPTER 5

Deepening Divisions

"I find that because of modern technological evolution and our global economy, and as a result of the great increase in population, our world has greatly changed: it has become much smaller. However, our perceptions have not evolved at the same pace; we continue to cling to old national demarcations and the old feelings of 'us' and 'them.'"

Dalai Lama

Since the invention of collaboration, we have sought structures and systems that foster and require unity or at a minimum conformance. From our nomadic ancestors' early hunting forays to when they began to settle in villages, they quickly learned that to survive they had to work together. They learned that working together required the guiding parameters of customs, rules, and laws. These "definitions" would ensure that their collective effort was less vulnerable to the errant, self-serving actions of the few or the warring tribe next door. In short order the cultural cohesion and uniformity motivated by their customs, rules, and laws were joined by other definitions of their emergent communities, namely beliefs, values, and principles, definitions often embodied in what we now call religion. From the beginning there was a recognition by humans that checks and balances mattered. Our ancestors realized that without a shared understanding of right and wrong, without defined consequences attached to non-conformance, without consensus agreement regarding the fundamental operating principles of a society, chaos and inequity would rear their heads, collaborations would fail, and human progress, their progress,

DOI: 10.1201/9781003089902-8

would likely come to a grinding halt. The Babylonian King Hammurabi clearly understood this, as captured in his preface to Hammurabi's Code, which is believed to be one of the oldest sets of laws ever crafted, sometime during his reign from 1792 to 1750 BC:

> Anu and Bel called by name me, Hammurabi, the exalted prince, who feared God, to bring about the rule of righteousness in the land, to destroy the wicked and the evil-doers; so that the strong should not harm the weak; so that I should rule over the black-headed people like Shamash, and enlighten the land, to further the well-being of mankind.

NEW RULES AMIDST A REPEATING PATTERN

For 4,000 years the progress of humankind and the rise of civilization were furthered mightily by the steady evolution of an enabling human code. In the last 30 years that code has been challenged by the unfettered, meteoric rise of technology and the appearance of an entirely new rule book. It is a book written not by our global village elders but by technology and its iconolastic innovators. The rule book's focus is not on the critical functions and behaviors of a healthy society and how to achieve meaningful human progress, but on how best to take advantage of human vulnerabilities and so-called needs to motivate universal adoption of the innovations being introduced. Concurrently, the unrelenting advance of the technology train has accidentally but decidedly undermined the humanistic concepts of shared benefit and shared responsibility. In doing so, it has weakened the power and role of the institutions that historically provided the trusted guidance and a non-partisan voice of reason to the citizenry as they struggled with foes both internal and external. But technology has not been the only invasive and divisive catalyst. It turns out that humankind has a history of collapsing in on itself because of itself, as proven by the rise and fall of every empire since the beginning of empires. The consistency of our hubris ultimately and repeatedly doing us in was starkly captured in a short treatise by a British mercenary and part-time academic named Sir John Glubb in the 1950s. In his 26-page *The Fate of Empires and Search for Survival*, Glubb revealed that every empire from the Incan to the British had a power period of almost the same duration: 250 years or so. He goes on to present a remarkable categorization of the six ages or stages all empires follow, six ages that mark their rise, their point of no return plateau and their inevitable fall (see Figure 5.1).

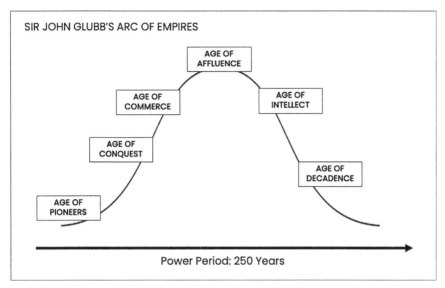

FIGURE 5.1   Sir John Glubb's Arc of Empires.

Not coincidentally these six ages/stages can just as easily be ascribed to many formations of humankind, from family dynasties to global corporations. And without hyperbole, we could now ascribe the six-stage trajectory to much of the Western world and America in particular. You will likely find Glubb's six stages strikingly and frighteningly familiar. They are applicable to where we have been, where we are, and where we will end up if we don't take matters into our own hands. We need the Humanist Revolution now.

THE WORLD ACCORDING TO GLUBB

Stage I: The Age of Pioneers. Glubb depicts the first stage as when a group of people (country, society, company) is singularly focused on carving out a better life, achieving a better existence or perhaps realizing a bold vision. His characteristics of "The Age of Pioneers" closely parallel that of entrepreneurs and their startups, reflecting an unwavering belief in the future, a willingness to sacrifice, to collaborate and compromise, to do whatever is required to advance the cause of the whole. Resources are limited but passion and conviction have no bounds. The effort required is so great that the people involved cannot or will not turn on each other. A focus on majority survival motivates unity. Along the way the dominant, perhaps natural, systems of hierarchies and patriarchies form. In all cases, the power and voice of the individual is supplanted by that of the structures that need to exist. Pioneers have always understood their place in the forming of society and in the universe. There is a shared understanding of progress and the explicit need for collaboration and compromise to achieve that end. This is how most startups start, how the Sumerians settled in Mesopotamia, and how America became the United States (although one could argue we went right to Stage II.)

Stage II: The Age of Conquest. As the pioneers make progress in establishing roots, they seek opportunities to grow faster by conquering other lands and peoples. The intentional acquisition of others brings with it both increases in resources and eventually complications due to a collision of belief systems, structures, and other differences, e.g., race and ethnicities. Even in the modern-day corporate version of this stage, call it "The Age of Mergers and Acquisitions", the long-term unintended consequences often end up being debilitating or even disastrous. A 2019 article in *Forbes Magazine* noted,

> According to the *Harvard Business Review*, between 70% and 90% of mergers and acquisitions (ultimately) fail. It's a shocking number, and the one thing all (conquests) have in common is people. Mergers and acquisitions fail often because key people leave, teams don't get along or demotivation sets into the company being acquired.

History has consistently shown that assimilation is not our strong suit. We tend not to get along over time with people not like us. It could be argued that conquest and assimilation have always been and still are inevitable vulnerabilities for any expanding entity or empire. The growing divide in the United States could be directly attributed to the growing diversity of our country. In the early days, the melting pot of America encouraged innovation and entrepreneurship, but a century later it fuels divide, unrest, and the singular concern of self-interest.

Stage III: The Age of Commerce. As the resources are gained and the acquisitive thirst abates, there comes a re-focusing on the basics of running the show, ensuring that the day-to-day economy is healthy and that most of the populace can meet their Maslovian base needs of food, shelter, safety, and belonging. America's "Age of Commerce" could be defined as the post-World War II period, when its massive war manufacturing capacity and carryover pioneer ingenuity was able to respond to the skyrocketing demand brought on by the re-building needs of much of the world. The "Age of Commerce" in America fueled the shared sense of the American Dream and the idea that any citizen could realize it. As some did, and some did not, tiny cracks began to form in the systems underpinning the society, with pockets of complacency, entitlement, and division appearing. The idea of the American Dream became a construct for individual accomplishment, individual freedoms and reward, not for collective progress.

Stage IV: The Age of Affluence. This is the stage of rapidly rising economic prosperity, with most of the citizenry enjoying big leaps in disposable income and the ability to move beyond their base level needs into the consumptive and increasingly addictive arena called "want." In this period, the aspirations of the individual begin to take precedence over the needs of the whole. Life is good for most, luxuries proliferate, and abundance increasingly becomes the general public's expectation. It is here that excess rears its ugly head. Systems begin to lose their collective humanistic orientation to fuel the economic growth that allows them to build their bureaucracies. The pioneer days of sacrifice, collaboration, and compromise are distant memories. As Glubb wrote, "Gradually, and almost imperceptibly, 'The Age of Affluence' silences the voice of duty. The object of the young and the ambitious is no longer fame, honour or service, but cash."

Stage V: The Age of Intellect. The fifth stage is the leading edge of the entity's downward spiral. The presumption of wealth leads to the growing distrust of structures and a collapse of the belief systems that served the pioneers so well. The importance of the collective is completely lost as the individual's opinions, beliefs, and wants take center stage. Regardless of the form of the entity (empire, country, or company) in the "Age of Intellect," doing less and just holding on becomes the standard operating procedure. Self-sacrifice is obliterated by self-promotion and an unwavering conviction that the entity's continued success and market dominance is guaranteed. This is the story of RIM (Research in Motion) and hundreds of other companies that were once market leaders. It is the story of every empire since the first empire, and it is the story of much of the Western world today. A story that is made more pronounced by technology.

Stage VI: The Age of Decadence. The final stage results from an extended period of wealth and with that a rapid expansion of delusional hubris.

Decadence, in Medieval Latin, means "decay" plus "to fall." "The Age of Decadence" is the time where the social fabric begins to rot and the divide between the wealthy and the poor becomes a gaping chasm. Myopia reaches epidemic status, isolationism becomes a proffered cure-all, materialism is the golden calf, and the entity's leadership tries to pull back from over-extending welfare supports, like universal health care. Big bubbles appear and often burst. In "The Age of Decadence", fear, greed, and pessimism emerge as the tenor of the times, and it appears as if the populace is getting more ignorant and selfish by the minute. The quest for personal development is replaced by an almost perverted focus on entertainment and hedonistic pleasure. Intellectuals become a pariah for many. As Glubb wrote in 1950, in reference to the then American and British cultures:

> (In *The Age of Decadence*) the heroes of declining nations are always the same—the athlete, the singer or the actor. The word 'celebrity' today is used to designate a comedian or a football player, not a statesman, a general, or a literary genius.

In Glubb's view, the past fall of Great Britain and the impending fall of America had and have a great deal in common, in large part because of our primal human tendencies. Regardless of the century or empire, it appears like we do this to ourselves. Technology has just accelerated the conclusion.

TECHNOLOGY AS AN ACCELERANT OF THE CURVE

Glubb's treatise is a concerning and yet useful view of the history of the empires and of humankind's tendency to allow economic success over time to foment division, individual rancor, and greed. With it comes a breakdown of the systems and structures that enabled the success to begin with. The depiction of all empires following the same up and down path is proof positive that certain human behaviors are both consistent and ultimately problematic. The fall of empires, the potential fall of the developed world, is not just a capitalistic or technological consequence, it's a humanistic one. With untethered economic growth and absence of any moral checks and balances, inequalities deepen and our individual needs and wants take over, eroding the foundation of shared values, shared outcomes, and shared means to achieve those outcomes. The growing concern about the tenuous state of democracies in the world reflects this erosive state.

What's different about today's developed world and its number one superpower, the American empire, is that the invasive, divisive force of certain human behaviors has for the first time been joined by the invasive, divisive force of technology. Arguably it is condensing and shortening the 250-year power period and making the divisions and dysfunction that much more complex and challenging to bridge. The advances of technology have exponentially magnified the separation between us, politically, economically, racially, morally, and ironically, digitally while representing profound contradictions that threaten our collective futures. We appear unable, even when faced with many an existential threat, to muster the collaboration and conformance required to mount an effective response. The gaps between us grow, as does the gap between technology and humans struggling to close it. One of technology's greatest unintended byproducts, our multi-modal, real-time capacity to get and share unfiltered information, has changed everything, and changed us, not for the better. The change in us is perverting our willingness to respect, reinforce, and refine the systems and structures that we need to keep us united and for many, subverting their desire to be united.

MISINFORMATION: THE GREAT DIVIDER

Without question, information is the lifeblood of communication and communication has always been central to how humankind works and its unique ability to collaborate. Many argue that it is Homo sapiens' evolved capacity to communicate that has most separated us from all other living creatures. Thanks to technology, the nature of human communication, individual and societal, is profoundly different than it was even 20 years ago. What we say, how we say it, what we see, what we hear and what we believe is less and less a rational, truth based, and expertly guided proposition. The age-old vetting filters and validating, curatorial hierarchies are largely gone. In the new order, the social networks have flattened the playing field, redistributed the power from accredited third-party institutions to the individual and, even more alarmingly, to a flotilla of trolls scattered around the world hell-bent on causing political and societal upheaval in a myriad of forms and countries. In a March 2020 article, *The New York Times* depicted one of the most organized and insidious trolling efforts, Russia's so-called Internet Research Group:

> The Kremlin-backed group was identified by American authorities
> as having interfered in the 2016 (Presidential) election. At the time,

Russians working for the group stole the identities of American citizens and spread incendiary messages on Facebook and other social media platforms to stoke discord on race, religion and other issues that were aimed at influencing voters.

As the technology-armed trolls enjoy free reign to create discord and division, thanks to the social network, anyone's opinion, no matter how unqualified or unsubstantiated, can now be shared in milliseconds with millions of people at essentially zero cost. In the United States this newfound unlimited distribution capacity puts a spotlight on the First and Second Amendments in our Constitution and how technology may have changed their relevance or at least might suggest some needed tweaks. The First Amendment, ratified in 1791, declared that all Americans had/have a right to say pretty much whatever they want to say without fear of government interference. That so-called Freedom of Speech amendment was written at a time when getting one's opinion heard required Herculean effort, from printing and hand distributing flyers to standing on soap boxes in public squares yelling at the top of one's lungs. The limits on distribution effectively served as governors on the integrity of the message being purveyed. Now that the distribution governors have been completely removed, so too has any ability to insure or confirm the integrity of what is being shared. Much harm can and is now being done by our words because technology has exponentially accelerated our ability to deliver them freely, challenging the ability of others to monitor and, if necessary, edit them. An October 2020 article in *The New Yorker* underscored the magnitude of the task:

> In Clegg's (a Facebook executive) recent blog post, he wrote that Facebook takes a "zero tolerance approach" to hate speech, but that, "with so much content posted every day, rooting out the hate is like looking for a needle in a haystack." This metaphor casts Zuckerberg (Facebook's CEO) as a hapless victim of fate: day after day, through no fault of his own, his haystack ends up mysteriously full of needles. A more honest metaphor would posit a powerful set of magnets at the center of the haystack—Facebook's algorithms, which attract and elevate whatever content is most highly charged.

The impact of technology on the Second Amendment, the freedom to bear arms, provides a telling comparison. In firearms as in speech, technology has changed the context of our freedom by increasing speed and access and eliminating the need for carefully honed skills or validated expertise. The Second Amendment of the United States Constitution, states "A well-regulated Militia, being necessary to the security of a free State, the right of the people to keep and bear arms, shall not be infringed." The author David Grace once wrote in a Medium post, "In 1791, it took the average gun owner about thirty seconds to load and fire his gun, and it took great skill and practice to be able to hit what he was aiming at." In 2020, the owner of an AK-15 semi-automatic rifle can fire approximately 300 rounds in that same 30 seconds, hitting a target over 100 yards away. And thanks to 31 out of America's 50 states not requiring a gun permit or a license, no training is necessary because at 600 rounds per minute, having decent aim is moot. The life-as-we-know-it threatening impact of these accidentally recalibrated freedoms is real. And while speech is directly responsible for far fewer deaths, the loss of quality controls and distribution governors carries deadly consequences for the health of civil discourse and by extension, democracy.

The act of speech, of communication, is always a two-way proposition, one requiring an audience. And thanks to technology, just as we can communicate with virtually anyone with the touch of a button, we can also choose to only hear the few that we want to hear from. With a single swipe or press, we can suppress opinions that differ from our own, ensuring that our views and beliefs never have to compromise with or consider those of others. The information sources we often choose increasingly carry little to no responsibility for providing supporting facts or corroborating evidence. Claims no longer require substantiation. Hard truths are often labeled as "fake news" and true fake news is no longer considered by many an intellectual or ethical offense. Lies are increasingly an accepted way of life, with fewer and fewer citizens and leaders taking responsibility for what they say and how they say it. As noted, in America, our most fundamental freedom, the freedom of speech, is revealing a vulnerable underbelly. It is an underbelly exposed and bloated by the social networks and the breakdown of our proofing structures and editing systems. Without the filters and fetters of truth, objectivity, facts, integrity, consideration, and compassion, the freedom to speak is increasingly becoming fuel for the fires of a growing and ugly vitriolic war that is further dividing much of the developed world. In his recently published encyclical entitled *Fratelli Tutti* or *All Brothers*, effectively a plea for global fraternity, Pope Francis declared this:

> Even as individuals maintain their comfortable consumerist isolation, they can choose a form of constant and febrile bonding

that encourages remarkable hostility, insults, abuse, defamation and verbal violence destructive of others, and this with a lack of restraint that could not exist in physical contact without tearing us apart. Social aggression has found unparalleled room for expansion through computers and mobile devices.

THE FEAR FACTOR

The fires of aggression and division are being fanned by two primal emotions: fear and selfishness, the causal emotions that Glubb implicitly cites in his depiction of every empire's final "Age of Decadence." In a talk given on nuclear weapons at the Atomic Bomb Hypocenter Park in Nagasaki, Japan in November 2019, Pope Francis underscored the dysfunction of fear:

> Our world is marked by a perverse dichotomy that tries to defend and ensure stability and peace through a false sense of security sustained by a mentality of fear and mistrust, one that ends up poisoning relationships between peoples and obstructing any form of dialogue.

The vitriol and partisanship being expressed across America and in many countries around the world are sourced from three overlapping sensibilities: loss of status, a sense of unfairness, and survival fear. The growing popularity of authoritative leaders and their populist and often protectionist policies are a result of the traditional majority population in many countries feeling that they are economically vulnerable, that their deficiency needs may not be met, and that the prosperity and opportunity that they once had, that they were entitled to, has been unfairly taken away. As Will Storr, the author of *The Status Game: On Social Position and How We Use It* convincingly proposes,

> This is why poverty alone doesn't tend to lead to revolutions.... (they) have been found to occur in middle-income countries more than the poorest. Sociologist Professor Jack Goldstone writes, 'what matters is that people feel they are losing their proper place in society for reasons that are not inevitable and not their fault'... . when we and our people sense our collective status is in decline, we become dangerously distressed.

The old status quo has been blown away by technology. Technology has globalized the playing field, eliminated entire industries, eradicated barriers to entry, and made survival of the capitalist fittest the order of the day. Along the way governments have neglected to explain the new order or provide the re-skilling and re-training the now fear-riddled traditional majority needed to maintain their socio-economic position. Lacking real understanding or the means of moving forward, the marginalized majority seeks the comfort of figureheads who will promise them a return to the old status quo past while not having to explain exactly how that might happen. Loss of status, a sense of unfairness, and survival fear are all sensibilities that obviate the need or interest in exploring the depths and dimensions of the problem at hand. Instead they serve as a free pass to the aggrieved to complain that they've been wronged with zero consideration of how exactly to make it right.

The slow erosion of both position and opportunity among the empire's (today's Western world) longstanding majorities have consistently translated into a resurgent fear of minorities, of people "not like them," i.e., of immigrants. It happened in Rome in 476 AD and it is happening today. Racism and its frequent companion nationalism are largely fear-based propositions, reflecting most humans' struggle to embrace other races and ethnicities that look different, act different, and hold different beliefs. As the world has globalized, as mobility has improved, as natural disasters have become more frequent, as genocide and civil wars continue to abound, immigration has skyrocketed. According to the United Nations, the number of international migrants topped 281 million in 2020, an increase of 60 million adults and children (almost 20%) since 2010, and representing 3.6 percent of the global population. Most of the movement has been towards the European Union and America, resulting in growing citizen backlash in those countries and severely tightened immigration restrictions.

Racism, protectionism, and the appeal of nationalism, are clearly not new. In our reptilian quest to survive we subconsciously seek to subjugate or prohibit the different, the foreign. We subconsciously seek to convince ourselves of our moral and intellectual superiority while we scheme to ensure that the outsider will forever be limited in their capacity for equality and the economic prosperity that accrues to it. In times of increasing uncertainty and the growing likelihood of personal setback, the majority of peoples' fears deepen as does their need to feel superior to others in order to combat the gnawing feeling of declining status. With that comes a worry that the minority will take what is rightfully theirs, as the longstanding majority, an entitlement of generations. Their vision of the future is no longer of the future but of the past, returning their country to "the good old days" when minorities would be occasionally seen but never heard and the majority's power and position was guaranteed. The obvious irony in the return to the past stance of the majority against the minorities is that in most cases, the majority came from the minorities. As an example, America is a country of immigrants ("The Age of Pioneers" in Glubb's six-stage progression) that conquered the indigenous people ("The Age of Conquest"). The immigrants represented a multitude of minorities and now, 250 years later ("The Age of Decadence"), they, who once were us, are the enemy. And we are afraid of them.

At the center of all this is the gaping and growing hole of economic inequality. The chasm between the haves and have nots in America has been dug deeper and wider by technology. Furthermore, the setbacks and economic challenges associated with COVID-19 have made the divisions that much more apparent. The digital marketplace and tech-centric landscape play to those with the greatest access to the data and the sophisticated tools that can turn that data into strategic advantage, greater wealth, and the capacity to exclude other people from achieving it. Technology has fundamentally turbo-charged the capitalist system in the United States, allowing the rich to get richer and effectively eroding the segment of our society that once embodied the American Dream, the middle class. According to a wealth distribution report issued by the U.S. Federal Reserve in 2021, the wealthiest top 1% of Americans held $36.2 trillion in net worth, surpassing the middle 60%, which held $35.7 trillion. The gap was the first time the 1% held more in the history of the report.

Once again, the pandemic further exposed these longstanding divides and, in some cases, made them worse. The families and individuals with wealth and latitude were able to respond to the challenges, while those who lived hand to mouth, paycheck to paycheck, struggled to keep up. As one example, a 2020 article in Axios described how some parents were addressing the loss of schooling time due to COVID-19 and school lock downs: "Neighbors are banding together to hire private instructors as a way to secure childcare and make up for some gaps online-only classes will leave in their kid's educations." These so-called pandemic pods cost as much as $1,000 per month per family, a cost that was and is out of reach for most families. The Axios report went on to talk about the inherent racial inequality of COVID-required online-only primary and secondary education:

> Students will likely lose, on average, 6.8 months of learning if in-class instruction does not resume until January 2021, according to an analysis by McKinsey. But Black students may fall behind by 10.3 months, Hispanic students by 9.2 months, and low-income students by 12.4 months.

Racial and economic inequality are being perpetuated and exacerbated in part by the capacity of some, but not all, to use technology and the financial ability to either respond to challenges or take advantage of opportunities. The decline of our understanding of us and our understanding and embrace of others, is fueling that division.

EXTREMISM ON THE RISE

As fear foments the divisions of color, cultures, and countries it serves as an accelerant for partisanship and the emergence of extreme political parties like the recently banned Neo-Nazi Nordadler faction in Germany, the Partido Social Liberal in Brazil, the White Aryan Resistance in America, and the anti-immigration group Britain First. Hundreds if not thousands of alt-right, alt-left, fascist, and white supremacist organizations have appeared over the last 20 years in large part because the extremists that belong to them can easily find each other and find ready fuel for their incendiary, radical positions. All of this is because of the now nanosecond ability to source and share information. Twenty years ago, it was difficult for a Neo-Nazi in Frankfurt, Germany to discover people who agreed with him or her in Frankfurt, let alone other parts of Germany. Now, a couple of clicks is all that keeps them apart. Alongside these rapidly organizing efforts to suppress minorities and paint those of opposing beliefs as archenemies, are a multitude of perpetuated, completely unsubstantiated conspiracy theories. They range from the long rumored Deep State in America, the theory that there is a hidden shadow government that runs America, to QAnon, a far-right and far out allegation that a group of Satanic pedophiles operating a global child-sex trafficking ring are somehow plotting against former President Donald Trump. That absurd theory goes on to propose that there will be a day of reckoning where large numbers of liberal politicians, celebrities, and members of the media will be arrested for their scheming ways and that Donald Trump will emerge as the savior of America. As foundationless as that proposition is, a Facebook study in 2019 identified millions of QAnon followers participating in thousands of pages and group postings. Thankfully, Facebook opted to bend their supposed commitment to Freedom of Speech and took many of those pages down. But that does not mean the extremism and fear mongering have been taken down too. In fact, in the twisted logic of the conspiracy theorists, being barred from sharing their racist, anarchist sentiments on Facebook translates into further proof of the importance of their work and the need to further fan insurrectionist flames.

Conspiracy theories abound, as starkly revealed in the results of 2019 survey of 1,220 U.S. adults by YouGov and Statista. 27% of Americans believe the U.S. government is hiding aliens in Area 51, a high security area in Nevada. 22% believe climate change is a hoax.

SELF-INTEREST VERSUS SELF-SACRIFICE

As the developed world increasingly divides along political and socio-economic lines, it is also seeing a growing separation between those who care about others and those who don't, another common attribute of every empire's "Age of Decadence." It is the time when most of the populace employs self-interest as its standard for decision-making, an orientation fueled by increasingly diverse populations and our hardwired biological and psychological distrust of people who don't look like us. In a 2014 article in the *National Review* titled "Homogeneity is Their Strength," author Kevin Williamson called it out, "We are more inclined to share and to cooperate with people to whom we are related, and we are most likely to trust faces that look like our own." He goes on to declare,

> How wide we draw the circle of kinship and how we think about its boundaries are cultural issues, true, but our habit of scrutinizing and categorizing, and of adapting our behavior accordingly, is as much a natural part of us as our blood and bones.

Again, in Will Storr's book *The Status Game*:

> Humans have a bias for their own that's universal, subconscious and triggered at the slightest provocation…in one study, 5-year-olds were given a coloured T-shirt to wear and were then shown

pictures of various children, some in matching shirts, others not. They knew their colour was randomless and meaningless and yet they still thought more positively of kids wearing the same colour, believing them to be more generous and kind.

The obvious and unfortunate flip side of this apparently universal truth is that we are less inclined to trust and share with people who are less like us. Fear and self-interest are intrinsically connected, and now thanks to technology-fueled globalization and misinformation flow, are leading to greater and greater disconnections. In his article's summation, Williamson shares a pointed statement made by Harvard's Alberto Alesina and Eliana La Ferrara of Bocconi University in their paper, "Ethnic Diversity and Economic Performance":

> The potential benefits of heterogeneity come from variety in pro-duction. The costs come from the inability to agree on common public goods and public policies.

While a difficult appraisal to read, the fact is that we cannot retreat from the growing diversity in the world. We must go forward to repair the divisions between us, not expand them. Humankind is humankind, and our capacity not just to thrive but to survive is predicated on more of us understanding that we are one, regardless of the color of our skin or our beliefs. A planet dominated by the selfish "me" versus the selfless "we" is destined to collective failure, the same failure that all Glubb's empires ultimately experienced. When the COVID-19 pandemic panic arrived in the United States in March 2020 there was a run on cleaning supplies, paper towels, and meat. Around the same time *The Wall Street Journal* published an article about the pandemic's impact on British grocery shoppers, writing,

> Early Sunday, in a joint letter published in British newspapers, the country's largest supermarkets asked shoppers to "be considerate in the way they shop" and stop buying more than needed. There is enough for everyone if we all work together, said the letter.

The operative word was, and is, if. But it should be joined by another word, when. For it is true that one day we will have to bridge our divides, to sacrifice to survive. The existential threats are not going away. The idea of the Humanist Revolution is to not wait for that Armageddon Day to motivate us to come together. Instead let us motivate our individual selves now to catalog our biases and unhealthy behaviors to begin the process of changing them, and in doing so, change the outcome for our species. Our behaviors got us into this mess, our behaviors will get us out.

LAGGING INSTITUTIONS

Division is becoming the most troubling manifestation of our increasingly fearful and egocentric selves and it has now exacerbated by an alarming erosion of the institutions and systems that have thus far enabled our progress. Across the developed nation landscape there is a growing and legitimate concern regarding the performance of our systems of administration, of justice, of law and order, of health care, of education, of commerce, and even democracy itself. There is a growing and deserved questioning of whether these longstanding systems can continue to deliver given the nature and demands of the world we now inhabit. The acceleration of technology is painfully exposing their limitations and underscoring how in today's world for any core system or structure to not just remain relevant but to perform effectively, it must innovate, it must adapt, and it must change, just like us.

Consider this fact: our systems and infrastructures were all built in and for a decidedly different world, an industrial age that operated at a far slower speed and presumed both steady growth and that the way it was, was largely the way it would always be. Subsequently, the gap between what the needs of modern society are and what our legacy institutions can deliver is growing wider and wider, at a pace that parallels the widening chasm between the technological advance and humankind's ability to keep up.

There is a growing, albeit slow acceptance of the need for our institutions to shift their operandi from remediation to prevention, from maintenance to innovation, and from status quo spending to greater investment in next generation capabilities. Our core systems will have to adopt the mindset and behaviors of startups, i.e., constantly testing, iterating, and improving their ability to meet new customer demands, withstand competition, and weather the onslaught of more nimble alternatives. America's decaying and lapsing infrastructure is exhibiting none of the above behaviors. Its bridges, highways, and ports are woefully underfunded and out of synch with the ever-increasing movement of goods. A 2017 study by the American Society of Civil Engineers reported that of the 614,387 bridges in the United States, approximately four in ten were 50 years or older, and almost 10% were structurally deficient, even while accommodating 188 million trips across them each day. Due to a lack of preventative maintenance, repair costs are currently estimated to be more than $200 billion. The good news is that with the recent passing of the Bipartisan Infrastructure and Investment Jobs Act, something is going to be done about it. The bad news is that the commitment is for $40 billion of new funding for bridge "repair, replacement, and rehabilitation", $160 billion shy of what is needed.

As the bridges are falling down, so too are our Internet speeds. In a 2017 piece written by Bhaskar Chakravorti, the Senior Associate Dean, International Business & Finance at Tufts University titled "*Is America's Digital Leadership on the Wane?*", he presents a compelling case that America's digital dominance is coming to an end, as we pull back on our (innovative and preventative) investments in infrastructure and our federal government shifts its technology focus from essential economic impetus to risk mitigation. When compared with its other OECD member countries, the United States, arguably the inventor of the World Wide Web, now offers its citizens some of the most expensive Internet access and some of the slowest speeds.

THE FUTURE INSTITUTION AND THE COMMON GOOD

The erosion of the developed world's structures and systems extends beyond the physical to the cultural, to belief and to their role in fostering the common good. More and more people today are questioning the role of governments, of capitalism, and of religion. The institutions that many have treated as the bedrock to civilization's progress are increasingly being challenged for their purpose and/or performance by both the old and young. In an October 2020 piece in *The New York Times*, Yuval Levin, director of Social, Cultural and Constitutional Studies at the American Enterprise Institute was quoted:

> Doubt about the trustworthiness of the country's governance has been an animating feature of the American right, most recently as a force behind the Tea Party movement in 2009. Trump followers and Tea Party supporters begin from the premise that the institutions are all corrupt and they turned against us. And that is the essence of populism that (more) politicians are going to continue to emulate.

It's a sentiment that extends across the world. As reported on by The Institute for Human Science, a Vienna, Austria-based independent research firm focused on the study of humanities and social science,

> Globalization and European integration, combined with new opportunities opened up by the digital revolution, have led to a radical questioning of the legitimacy of the institutions of representative democracy, and have sharpened tensions between national democracies and the global market, and between the principles of democratic majoritarianism and those of liberal constitutional-ism.

People today are willing to question the integrity of how the world works because they can. They know, intuitively or not, that the societal systems largely designed decades ago have not kept up with the complexities brought on by the accelerating technology train. Technology has exposed the disconnects, the misalignment, and the shortcomings of the legacy structures that were built to help us. It has also exposed the shortcomings of us, shortcomings perhaps fueled by the loss of human purpose among the systems that we rely on.

There is a convincing view that as 20th century society, increasingly fueled and led by technology, deepened its embrace of the sciences and the power of capitalism to drive economic growth, that performative individualism overtook collective well-being as the purpose of the systems and the state itself. Rather than the systems of governance, politics, economy, and education reinforcing the importance of and contributing to the common good through position and policy, and the shared morality and ethics required, they stepped out of the conversation to focus solely on functional progress and the game of who is ahead of whom. The loss of the human thread, the shared concern for a moral and ethical good, among our core institutions is a critical erosion of modern life, particularly in the case of higher education. As noted by Bellah in *The Good Society*:

> The very diversity of American education allows a variety of forms that would link intellect with character and citizenship. For these to flourish we must make changes throughout our institutional life, particularly in our economic and governmental institutions, changes that would show we understand education less obsessively

in terms of "infrastructure for competition" and more as an invaluable resource in the search for the common good.

The story of humankind has proven beyond a doubt that for any entity (empire or not) to survive let alone thrive its participants must be largely aligned and enabled by systems of law, order, principles, and beliefs that most of the population subscribe to and benefit from. Mirroring Glubb's "Age of Decadence," the empire of the developed world is an empire increasingly divided, supported by institutions that should be leading us but instead struggle to keep up. The combination of human division and structural erosion both real and imagined is producing a dark void that threatens to quickly deepen our time in the "Age of Decadence" and result in a chasm we can no longer cross and a point of no return for humankind.

It is a chasm we must bridge. We cannot realize continued collective human progress without returning to our collaborative, humanly connected roots. We must find our way to shared intentions, shared means and respected, adaptive and effective supporting structures and systems. These are the foundations for the coming Humanist Revolution. To realize these outcomes, we will need to first embrace the fact that, just as technology and humanity have unintentionally joined forces to turn us against each other and our ways of being, we can marshal those same capacities to bring us back together. The key is putting the truth of our humanity first, intentionally recognizing and sequestering our fearful and selfish tendencies, while we elevate and celebrate the importance of the whole and each other. With that we must be willing to let go of our legacy systems. We must replace them with human-centric, synergistic institutions that embrace technology as a tool for change and gauge performance through the right measures of human progress. The critical question remaining is where the leadership for these changes will come from. All indicators suggest it will need to come from us.

An invasive vine, its delicate yet voracious tentacles have insinuated into every nook and crevice of our structures and systems causing both the degradation and transformation of our ways of being and the institutions that have enabled us.

CHAPTER 6

The Shifting Sands of Power

"I think it only makes sense to seek out and identify structures of authority, hierarchy, and domination in every aspect of life, and to challenge them; unless a justification for them can be given, they are illegitimate, and should be dismantled, to increase the scope of human freedom."

Noam Chomsky

Another major unintended consequence of technology's acceleration past our human truths and institutions involves the matter of power. As we humans are wired for war and fixated on status and standing, so too are we persistent in our search for and defense of power. Power is both the great enabler of our progress and perhaps the greatest threat. It is also one of the most amorphous assets we can acquire, as perfectly captured by Leo Tolstoy, the author of *War and Peace,* in which he wrote, "Power is a word the meaning of which we do not understand." As true as that may be, it is also true that whoever holds the power and how it is deployed determines the fate of empires, the future of the environment we rely upon and our species. The story of humankind is effectively a story of power gained, power lost, and the consequences and costs of both. And thanks in part to the unrelenting advances of technology the power stakes are now existential.

Hierarchy has long served as the primary structure of power, a stratification exhibited throughout nature and manifest in every part of human society, from governing models and corporations to family units and psychological frameworks. The latter is represented in Maslow's Hierarchy

DOI: 10.1201/9781003089902-9

of Needs with its inference being that the top is hard to reach and its achievement the ultimate state.

Readily accepted "pecking orders," explicit dominance by the few, and the willingness of the majority to be led appear to be preordained structural constants that enable any group to best meet its Maslow-codified base level needs of survival, safety, and belonging.

HIERARCHIES AND GOVERNANCE

A look back in time reveals a prehistoric adoption of hierarchical formations beginning as early as 10000 B.C. with the coming together of the first farming communities. Human hierarchy research conducted by Professors Peter Turchin of the University of Connecticut and Sergey Gavrilets of the University of Tennessee points to a simple rationale: as people settled together for the first time and the size of settlements surpassed 150 people (the Dunbar-effect-calculated maximum number of manageable relationships), the early settlers responded to the looming chaos by establishing clarifying tiered hierarchical structures to "run" the community. The resultant organizing framework yielded exactly what was sought: decreased risk, explicit interdependencies, collective operating efficiency, and, by unintended extension, fealty for the person at the top of the pyramid, in most cases, a chief, warlord, or king.

The ultimate hierarchy, monarchies, quickly became the dominant power model across the world, until the 5th century B.C. when some radical thinkers in Ancient Greece introduced the modern concept of democratic rule. Their proposal that governing power should accrue to the masses had imminent appeal (but only if a country's economy was flourishing and the nation was not at war, which was rarely the case). The inflection point for the subsequent manifestations of democracy (and perhaps a foreshadowing of the fate of modern American democracy) was the Roman Republic's attempt to install it as a governing system. The complexities and diversity of Rome's sprawling, uneasy collection of tribes, the never-ending conquests, and the insatiable power grabbing of its leaders, all but guaranteed the failure of the Roman democratic experiment. The rule of the people was quickly replaced by the rule of the few and eventually the rule of one and the republic became an empire led by a string of autocrats known as emperors. We were back where we started. An attempt to distribute the power structure had resulted in the exact opposite effect, a hyper-hierarchy. The Western world's democratic experiment was effectively put on hold.

Centuries later a movement towards monarchical absolutism, pithily captured in the French King Louis XIV's infamous declaration, "L'etat c'est moi," triggered the 18th century revolutions in Europe and America that gave democracy a second shot as the preferred model of governance. As more democracies began to take hold, it would have been reasonable to assume an eventual universal adoption of democratic systems and an outright final rejection of monarchical, authoritative regimes as supported forms of governance. The people would or should always prefer having the power, or at least some of it. But as modern civilization has unfolded, the universal adoption of democratic systems has not been realized, and the sands and sandcastles of power have continued to shift and re-form. The foundation of democratic systems has continued to be prodded and poked, in a quest to better match the nature of rule with the reality of the day and the changing needs and expectations of the ruled. Or perhaps the ebb and flow of democracies and the persistence of autocracies have simply been and continue to be the consequence of the inevitable desire of a few to retain or gain more power over the many. Power pays for those who gain it, and for some, power clearly continues to be all.

THE DEMOCRACY-AUTOCRACY DANCE

The multitude of players jockeying for position to influence and guide the trajectory and governing structure of the world is rapidly changing. Old power is colliding with new power, old hierarchies are having to accommodate more distributed models, and the traditional nation state architects and engineers are having to accept that the road forward is no longer theirs alone. Add to that the likelihood that China, a Communistic, autocratic nation that has long eschewed the concept of multilateralism will someday soon secure its place as the world's number one superpower. A recent *New York Times* article titled "An Alliance of Autocracies? China Wants to Lead a New World Order" underscores the central power dynamics at play:

> Mr. Biden made that clear in his first presidential news conference on Thursday, presenting a foreign policy based on geopolitical competition between models of governance. He compared Mr. Xi (China's President) to the Russian president, Vladimir V. Putin, "who thinks that autocracy is the wave of the future and democracy can't function" in "an ever-complex world." He later called the challenge "a battle between the utility of democracies in the 21st century and autocracies."

According to a 2019 research study entitled "*The Varieties of Democracy Project*," there were just 25 democracies and 131 autocracies ruling the world in 1950. Fifty years later it was an even split, with 87 countries theoretically ruled by the people for the people, and 89 still serving as sandboxes for dictators and self-declared leaders for life. But in retrospect this apparent 20th century trend towards more democracies was a false positive. The most recent edition of the Democracy Index, an annual survey conducted by *The Economist Intelligence Unit* found that of the planet's 167 countries and the roughly 8 billion people who live in them, just 8.4% experience what the Index labels a "full democracy." The power trend line is explicitly away from the rule by the people towards the autocratic rule of the people. In fact, the Index also shows that over a third of the world's population now lives under an authoritarian regime.

Another view, depicted in Figure 6.1, shows the up and down dance of democracies versus autocracies over time.

The surprising staying and re-appearing power of autocratic governments and the many failed attempts at democratic rule over time can be attributed to a multitude of factors (along with the ability of autocrats to exploit them). Autocracies are often born or enabled by weaknesses in constitutional underpinnings, the might of militaries, the inabilities of incumbent leaders to reach democratic compromise, and the vulnerabilities of

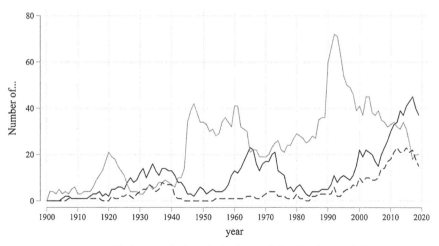

FIGURE 6.1   Democracy versus autocracy trends.

citizenry when unrest and uncertainty appear. Strong arms and dictatorial rule are sought or at least enabled by some segments of a populace when there are increased risks to their ability to get basic needs met. The current alarming rise of populism, nationalism, protectionism, and even fascism around the world coupled with the growing appeal of a singular authority are most likely a direct consequence of the rise of perceived personal risk, and with that, fear. Yes, we are back to fear. As the complexities and challenges of a global economy grow, extrinsic and intrinsic risks will only increase, and with that people will increasingly seek the comfort of strong voices, reassuring promises, and the opportunity to band together with their like-minded peers.

A NEW POWER SOURCE

From the Stone Age to the Industrial Age, power held was primarily the result of two things: hierarchical position and resources. Both were variables that are necessarily finite in supply, thereby limiting the number and type of power players. The transition to the Information Age that began in the 1970s and was fueled by exponential advances in technology (remember Moore's Law), heralded information as a new type of power source, reflecting a realization that when information is unbound, knowledge becomes the ultimate weapon in the war for more. And as the prolific spy master author Tom Clancy captured, it is a weapon that is not always used for good:

> The control of information is something the elite always does, particularly in a despotic form of government. Information, knowledge, is power. If you can control information, you can control people.

The democratization of information means that for the first time the traditional global hierarchy of power and its dominance by developed nations and their Western governments is under siege. The ongoing technology-fueled recalibration with information as the primary power source is causing a gradual but profound transformation of the global power structure. The result may be the ultimate unintended consequence of technological advance. The question of who or what might ultimately hold the power and whether or how it is deployed against the multitude of existential threats we face will be the central question of our time. And we cannot wait for the world's superpowers to answer it.

Nor can we assume that the democratization of information portends a strengthening of democracies. In a recent article written by Andrea Kendall-Taylor, Erica Frantz, and Joseph Wright for *Foreign Affairs*, the authors point out that contrary to original expectations, technology and information control are enabling autocrats to further secure their positions:

> Led by China, today's digital autocracies are using technology—the Internet, social media, AI—to supercharge long-standing authoritarian survival tactics. They are harnessing a new arsenal of digital tools to counteract what has become the most significant threat to the typical authoritarian regime today: the physical, human force of mass antigovernment protests. As a result, digital autocracies have grown far more durable than their pre-tech predecessors and their less technologically savvy peers. In contrast to what technology optimists envisioned at the dawn of the millennium, autocracies are benefiting from the Internet and other new technologies, not falling victim to them.

AN ADOLESCENT AGE

The power structure of the 21st century is now in an awkward transition, akin to a teenager struggling with the early stages of adolescence. With no parent as its guide or multilateral entity as its architect, it is clumsily experimenting with different degrees of centralized and decentralized power. The structure of our global society is jerkily transforming in some ways from a hardened and closed Western-led hierarchy of governments, a discrete system that the average citizen is subjected to, to a multi-polar, amorphous network that might one day become fully distributed, flat, and user empowered. Going or gone is the perhaps false comfort of an American government-led, Western-biased world order, a white, democracy-leaning superpower at the top of the global power pyramid. Effectively the new power structure that is emerging seems to be mirroring the information source of the power itself. It is allowing a multitude of voices to be heard and forcing a recognition of the fact that everything and everybody are now connected. While democracies as forms of government will likely continue to struggle to take permanent hold, and autocracies will seek to further secure their positions, it is still possible that the democratization of rule through a fully distributed system that enables a variety of players to participate, is on the horizon.

As clumsy and unclear as the transformation and transfer of global power control is, it is being made even murkier by the fact that the old rules of the information game have been replaced with new rules, rules that allow misinformation and disinformation to be parlayed and promulgated to nation-state-ruler advantage. The embrace of half-truths and non-truths by leaders and populace alike jeopardizes the viability of democratic systems by reducing policy debate and determination to a contest of unsubstantiated vitriol. When leaders can lie directly to other leaders and to their citizenry and avoid accountability, the existing world order is exposed, and the autocrat's welcome mat is extended. But this moment of such great uncertainty may be temporary, as more of us step forward to demand a reset of the rules and show a willingness to participate in the allocation and application of different types of power for the collective good.

THE EMERGENCE OF NEW POWER

In their book *New Power*, Jeremy Heimans and Henry Timms delineate the difference between "old power," the power associated with traditional, top-down hierarchies and "new power," the capacities and attributes that accrue to what appears to be the likely replacement. They write,

> Old power works like a currency. It is held by few. Once gained, it is jealously guarded, and the powerful have a substantial store of it to spend. It is closed, inaccessible, and leader driven. It downloads, and it captures. New power operates differently, like a current. It is made by many. It is open, participatory, and peer driven. It uploads, and it distributes. Like water or electricity, it's most forceful when it surges. The goal with new power is not to hoard it but to channel it.

For the first time in the history of the world, influence is beginning to come not from an empire or a rarified club of governments, but from an ever-expanding, loosely forming, and constantly shifting mélange of countries, corporations, NGOs, and even individuals who are all grabbing the microphone at different turns in the global progress road. As Heimans and

Timms explain, it's a transition of extremes, from a power model based on decree to one based on dialogue, from push to participation, from formal to informal, from controlled to untethered. The inference is that a flatter and more distributed power structure that matches the information network that fuels it will be a more effective one. But the authors also present a clear caution:

> The new power crowd would not have invented the United Nations, for instance; rather, it gravitates toward the view that big social problems can be solved without state action or bureaucracy. Often encountered in Silicon Valley, this ethos has at its core a deep and sometimes naïve faith in the power of innovation and networks to provide public goods traditionally supplied by government or big institutions. Formal representation is deprioritized; new power is more flash mob and less General Assembly.

DIFFERENT POWER PLAYERS AND PLAYS

As of today, we stand on a bridge between these two very different models of global governance. On one side is the old power, the old structure with some new, or not so new, players vying for the top power positions. On the other side is the new power, a diverse, amorphous collection of entities with their capabilities, ideologies, and perspectives tied together by the belief that the world is a network that should be run as a network by a network. Although this eventual transition from a closed power hierarchy to open network seems plausible, the ultimate end structure is impossible to predict. As Heimans and Timms point out, there are countries, companies, and organizations that are old power structures with new power sensibilities and there are new power structures that are already exhibiting old power structure behaviors. The latter suggests that whatever we do, humans will always struggle to completely remove hierarchy from the power equation.

The uncertainty that exists is made even more so by the advent of what Harvard Professor Joseph S. Nye, in his 1990 book, *Bound to Lead: The Changing Nature of American Power*, dubbed "soft power." At the time Nye's concept of soft power was seen as a radical declaration that the new world order was being formed and informed by the deployment not just of capital but of co-optation. He argued that as connectivity was on the rise, so too was the ability of influence (versus coercion) to impact nation state decision-making. Fast forward to today and soft power is being deployed by a wide range of players to varying effect, adding to the complicated picture. It also prompts a critical question. If a structured (and generally accepted) power hierarchy of national governments struggles to manage the growing complexities in the world, how well will an amorphous, undefined network without explicit leadership (and hard power) perform? And how well will it respond to the relentless advance of an army of existential and unvanquishable enemies?

China, parts of Asia, and select countries in the Middle East are the leading examples of old power structures with new power sensibilities and proof positive that the now centuries-old Western-led power hierarchy of global governance is coming to an end. America and most of the Western world are decidedly in Sir John Glubb's "Age of Decadence," manifesting much of the dissonance and dysfunction that tend to come at the end of an empire's peak power period. Countries and regions like China, the Asian Tigers, and the BRIC (Brazil, Russia, India, and China) represent the next generation of old structures with new power. They are all operating in the "Age of Pioneers" mode with an eye towards the "Age of Conquest." Their focus is on the acquisition and application of hard power through technological and infrastructural investment and soft power through strategic influence. These shifting sands of power will only shift more as predicted in the World Economic Forum's 2020 Global Risk Report:

> Today's emerging economies are expected to comprise six of the world's seven largest economies by 2050. Rising powers are already investing more in projecting influence around the world. And digital technologies are redefining what it means to exert global power. As these trends are unfolding, a shift in mindset is also taking place among some stakeholders—from multilateral to unilateral and from cooperative to competitive. The resulting geopolitical turbulence is one of unpredictability about who is leading, who are allies, and who will end up the winners and losers.

As it acknowledges the geopolitical turbulence and unpredictability the report also calls for all institutions to develop greater adaptive capacity:

As the outlines of the next geopolitical era start to emerge, there is still uncertainty about where the distribution of power will settle and from where influence will emanate, but a snap back to the old order appears unlikely. If stakeholders attempt to bide their time, waiting for the old system to return, they will be ill-prepared for what lies ahead and may miss the point at which key challenges—economic, societal, technological or environmental—can be addressed. Instead, longstanding institutions must adapt to the present and be upgraded or reimagined for the future.

CHINA'S SUPERPOWER GAME PLAN

While the longstanding Western-dominated citadels of power struggle to adapt and shift, China methodically marches forward, exiting the "Age of Commerce" towards "The Age of Affluence" (and influence), a transition that will likely bring it even greater economic growth but also more direct challenges to its Communist sensibilities and centralized power policies. A country of 1.4 billion, China's clear emergence as the imminent successor to America as the world's number one superpower can be primarily attributed to two things: economic reforms and technology. Just 40 years ago it was a decidedly backwater nation, largely disconnected from the rest of the world, with hundreds of millions of its people living in poverty. Forty years later, China's GDP growth (until very recently) is the envy of the world and is predicted by the CEBR World Economic League to surpass the United States in 2028 (see Figure 6.2).

According to the World Economic Forum, the once economic laggard has realized "the fastest sustained expansion by a major economy in history." The implementation of free-market reforms was effectively turbo-charged by technologies enabling a globally connected marketplace, a digital agora with billions of buyers and sellers, unleashing ravenous consumers who quickly came to expect infinite choices, lower costs, and access to anything they wanted when they wanted it, all of which were demands that could only be met by China's remarkable capacity to supply cheaply and quickly. It is that unique ability (and cost structure) that has, perhaps ironically, propelled an avowed Communist country to the top of the global capitalist power hierarchy. And in doing so, given them the motivation and ability to attempt to lock in their dominance by hoarding information at home and deploying technology and capital abroad. Globally, China is globally poised

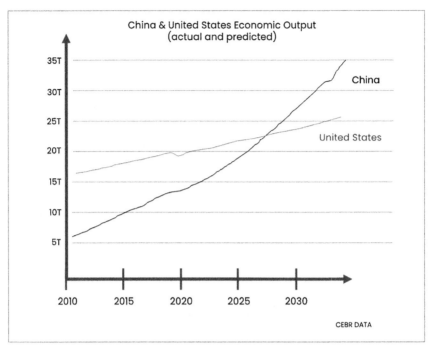

FIGURE 6.2   China versus United States actual and predicted GDP performance.

to take the lead on a range of fronts that were once assumed to be America's sole domain.

Whether a knight in shining armor or a calculating marauder, China's voracious quest for greater global presence and power is manifesting in many forms. As Hal Brands and Jake Sullivan wrote in a 2020 *Foreign Policy* magazine article entitled, "China Has Two Paths to Global Domination":

> There is the (Chinese) naval shipbuilding program, which put more vessels to sea between 2014 and 2018 than the total number of ships in the German, Indian, Spanish, and British navies combined. There is Beijing's bid to dominate high-tech industries that will determine the future distribution of economic and military power. There is the campaign to control the crucial waterways off China's coast, as well as reported plans to create a chain of bases and logistical facilities farther afield. There are the systematic efforts to refine methods of converting economic influence into economic coercion throughout the Asia-Pacific and beyond.

While the debate continues regarding the scope of China's geographic power intentions, the technologies that helped them become an economic power are going to power them to the unassailable top of the global power pyramid. Their audacious Belt and Road Initiative and its subset, the Digital Silk Road Initiative, look to embed China's interests throughout the developed and developing world via massive investments in land, sea, and, most importantly, digital infrastructures. In a recent opinion piece by Sam Olsen for *The Hill*, he calls out China's quest for digital hegemony:

> Later this year, we expect the release of the "China Standards 2035" plan, which aims to set global standards for evolving technologies such as the internet of things, artificial intelligence (AI), and 5G over the next 15 years. With Chinese technology infrastructure dominating in so many countries, the 2035 plan will cement China's standards as the norm and give its companies a significant, and perhaps permanent, business advantage over their American competitors.

Olsen's commentary also portends the troubling complexities of what happens when the world's greatest superpower is a Communist country,

> …(it) is a stark reminder of how technology is not an ethics-neutral domain, but instead is underpinned by subjective values that can be challenged. This is the real issue that Chinese technological control creates. Beijing wants to define the standards of important future technologies such as AI, and the values upon which they are based—a move that will tilt the world away from American commercial and political influence.

China is the decided head of the roaring Asian tiger, and seemingly hell bent on leveraging its economic and technological muscles to get to the top of the power pyramid, even while much of the world seems to be wandering across the bridge towards a different model, one that is far more open source and integrative. And that raises two critical questions: What happens to the world when its leading superpower is an authoritarian government that has shown a consistent track record of trampling human rights, of rejecting multilateralism, and ignoring the needs and concerns of the planet as a whole? And how does the old power structure with this new oppressive superpower at the top deal with the emergence of new forms and pockets of power that may be less willing to pay homage to governments as the arbiters of rule? The sands of power will no doubt continue to shift, as few if any players are willing or able to stand still, including Africa.

EMERGING AND ESSENTIAL POWER CANDIDATES

Several countries on the African continent are case studies of an entity leap-frogging old power structures and adopting new power sensibilities, subtly but clearly inserting itself into the global economic game and showing the world what happens when you unleash the power of youth amid an environment of constrained resources. The vast continent was arguably the birth-place of the first entrepreneur, the first pioneer, modern man, *Homo sapiens*. Today it exudes that sensibility in the form of new power. The culture of the continent represents a way of thinking and operating that effectively contradicts the ways of the old power structures. It is an approach that is focused on practical need, accepts limited resources, and is not tied to legacy infrastructures, while recognizing that for material progress to occur it takes the entire village. We can hope that many and more countries in Africa have embraced power as power distributed, power connected, power as means, not ends, and will emerge as a model for decentralized new power success.

While parts of Africa power up in new power ways and China aggressively marches to the top of the old power pyramid, multinational corporations are increasingly insinuating themselves into the task of global governance. It is a move enabled in part by the fact that they are increasingly seen by most citizenry as the last trustworthy player standing. The 2021 Edelman Trust Barometer, the long running gauge of what brands and entities consumer trust the most, revealed a growing distrust of governments. It also captured a growing perception that businesses alone may be capable of dealing with the world's complex challenges in an ethical, bi-partisan, and humanistic manner. The trust factor, joined by globalization driven by technology, has rendered the multinationals as a high impact node in the new power network while also transforming them into foot soldiers in nation-states' soft power and economic conquest, a development captured in this comment in 2018 on *The Conversation*:

> It starts to become apparent that international relations are any-thing but a one-sided story of either state or corporate power. Globalization has changed the rules of the game, empowering corporations but bringing back state power through new

transnational state-corporate relations. International relations have become a giant three-dimensional chess game with states and corporations as intertwined actors.

Whether a three-dimensional chess game or an open, dynamic network of nodes and connectors, the emerging new global power structure is far more complex and multivariate than the old. And while multinationals in general are playing a bigger and bigger role in how the game is being played, like the old power structure, a few are at the top, wielding incredible influence on billions of people lives and de facto the countries within which they live. There are five corporations that dominate the Western world and increasingly large swaths of the rest. And all are, unsurprisingly, technology companies. Originally labeled as FAAMG by the investment firm Goldman Sachs, Facebook (now META), Amazon, Apple, Microsoft and Google (now Alphabet) had a combined market capitalization in 2023 of over $7 trillion and represented almost 20% of the S&P 500 Index, one of America's leading measures of big company performance. But the FAAMG's financial valuation is irrelevant when compared with the role these five behemoths play as critical functions in the now essential task of finding, using, managing, and moving information. With information becoming the primary source of power, these five companies have become both the greatest enablers and by extension the greatest power holders. That capacity has moved them into the center of conversation regarding the foundational human questions of freedom of speech, equality, privacy, and the threat of monopolies, or loss of freedom of choice. But their role as critical economic drivers has the American government and others gingerly contemplating how best to both rein in their power while not undermining the economic momentum these businesses have created. As these five companies are prompting constitutional questions both domestically and globally, they are also increasingly replacing the U.S. federal government as the leading force behind the development of the next generation of America's foundation infrastructure, a trend captured in 2017 in an interview with *The New York Times* technology columnist Farhad Manjoo on NPR's *Fresh Air* program:

> …many of the technologies that we use today at kind of their earliest levels were started by grants for the Defense Department or just kind of basic science research. Now a lot of that is being done by these (FAAMG) companies. Artificial intelligence is kind of the primary example. These companies are going to be building

the future of transportation in the United States, in the world. You know, they're building self-driving cars. They're building drones. They're building kind of the infrastructure of the United States— the infrastructure of the next 20, 30, 40 years in ways that we used to look to kind of governments to do.

Like China, Singapore, and their peer "Age of Pioneer" and "Age of Conquest" countries, the FAAMG are fundamental old (hierarchical) power structures attempting to embrace new power sensibilities, gaining remarkable ground in their ability to steer the world while they deploy both hard and soft power. The emergence of unrestricted information flow has effectively eroded the walls protecting government-controlled messaging and policy formation in much of the developed world, opening up the populace to new voices and to more expansive consideration of how to define both societal and national problems and deliver potential solutions. In his 2004 book, *The Future of Power* and subsequent articles, Harvard Professor Nye dug deeper into the meaning and the means of soft power, particularly as wielded by NGOs. In one 2004 piece he wrote,

> Many NGO's claim to act as a "global conscience," representing broad public interests beyond the purview of individual states. They develop new norms by directly pressing governments and businesses to change policies, and indirectly by altering public perceptions of what governments and firms should do. NGO's do not have coercive "hard" power, but they often enjoy considerable "soft" power—the ability to get the outcomes they want through attraction rather than compulsion. Because they attract followers, governments must take them into account both as allies and adversaries.

According to *The Global Journal*, in 2022 there were 10 million NGOs in the world, led by longstanding players like Greenpeace and Oxfam, and focused on a range of cross-border problems from child malnutrition to environmental degradation. As importantly those 10 million NGOs were financially supported by over 2.5 billion individuals across the planet, reflecting a growing recognition that our planetary challenges require more than governmental intervention to solve, and that soft power is worth investing in.

The capacity and inclination to wield soft power extends to yet another type of player in the emerging distributed power structure, individuals. As information access has democratized, the influence of celebrity on the global game plan has grown exponentially. The Internet has become a sort of bully pulpit, allowing these players to express their views and beliefs in ways that are impacting the arc of modern society, influencing how governments govern, and how people live their lives. The list of soft power players is long and diverse, ranging from the environmental activist Greta Thunberg, philanthropists Bill and Melinda Gates, Jack Ma, the founder of Alibaba, and Mike Bloomberg of Bloomberg Media. The poster child for the power of an individual to sway the planet may be Elon Musk. The founder of SpaceX, the product architect of Tesla, the founder of The Boring Company, a startup that is inventing ultra-high-speed rail, the co-founder of SolarCity, and a co-founder of OpenAI and Neuralink, his influence is as much a consequence of the global following he has built and the power that that association produces as it is his commercial enterprises. As technological innovation has been Musk's business focus, it has also been the power behind his own soft power throne.

A MANDATE FOR DIFFERENT GLOBAL GOVERNANCE

The complexities of the emerging power model are surpassed only by the complexities of the myriad of challenges facing humankind. Our issues and enemies are daunting, the level of threat unprecedented, and the means to solving the problems unclear. The task at hand suggests that there is a need in the near term for a global governance structure that is more hierarchical than less. The "entity" must be carefully balanced, adaptive, multi-lateral and reflective of an agreed-to power order to enable efficient decision-making, global policy formation, and unilateral enforcement. Without that, at a minimum, an increasingly connected, interdependent world with ever-compounding problems requires greater coordination among the players and a more explicit definition and acceptance of how they need to work together to combat our many existential threats.

Professor Roland Paris of the University of Ottawa wrote in an article for *Ethics & International Affairs*:

> No amount of institutional proliferation or innovation can ultimately substitute for a lack of consensus among incumbent or rising powers on the fundamental "rules of the game" in world affairs,

including basic norms of political legitimacy, war and peace, and commerce. These are the foundations upon which any workable global governance system must be built.

And yet we appear headed in the opposite direction.

TECHNOLOGY AS POWER

As concerning as that picture may be there is something more bothersome, and that is the existence of another emergent power player rarely recognized, a power player that may carry the greatest power and influence of them all. The player is technology itself, and the technologists that are behind it. Intentional or not, they have been and continue to be focused on creating systems that permeate every aspect of society, systems that are frictionless, hyper-efficient, and fundamentally capable of eliminating steps and tasks that get in the way of speed, access, and economic growth. In that obsessive quest, they have unintentionally been creating the greatest risk, the greatest threat of all, the dehumanization of our existence. It is possible that the shifting sands of power will accidentally render a world where technology is both the sole means and the ends, a world where humankind has been left behind and in our stead is a machine that runs so efficiently that we end up not as beneficiaries but as its nominal cogs. In her brilliant critique of Brett Frischmann and Evan Selinger's book *Re-Engineering Humanity*, Laura Drake succinctly summarizes the problem:

> Whereas systems were once engineered to fit the priorities of humans, Frischmann and Selinger tell us that humans are now being programmed to fit the priorities of systems. They claim that what makes this 21st century version of "techno-social engineering," or "engineered determinism," as they alternatively call it, different from humanity's earlier encounters with technology is its scale, scope, influence, architectural extension, and the new factor of media intervention. These all add up to unprecedented power. We have now entered an era in which humanity is being overpowered by the systems it made.

The technology train continues to hurtle forward, effectively overwhelming the past and present while unintentionally fomenting a chaotic future of rule. It is a future that requires rewriting the rule book to serve humanity not technology; to serve our planet and not just ourselves. The likely near-term transition before us, from America to China-led, from democracies to autocracies, from old power to new power, from closed hierarchies to open networks, is destined to be troubled, full of tantrums, outbursts, and the occasional fender bender that all adolescent drivers are prone to produce. As technology is enabling and even forcing that transition, it is also participating in it as both the ultimate power, the ultimate threat, and ironically, the only likely means to solve the myriad of problems we face. For us to regain human depth and dimension in our lives, eschew our thirst for excess, repair our divisions, and combat the many specters of threat, including technology itself, we will need to step to the fore first as humanists. People who filter our decisions and structure our global society and its rule in ways that truly serve more humankind for decades and centuries to come. That essential action requires an intimate understanding of what it means to be human and how we can insert that truth into everything we seek, do, and create for ourselves and each other. This is what the Humanist Revolution is all about.

We have the power to take the power from technology, to establish the essential guard rails for our careful creation and adoption of technologies that serve humankind now and generations to come. With our new rule book in hand, we will then be able to become technology's steward instead of its servant. We first must agree on what our desired destination is, what serves our future and the futures of generations to come best. We have built an engine of prosperity. We need to apply it to the right purpose with the right zeal, and to do so together.

We must rebuild the bridges to our humanity.

III

The Humanist Revolution

CHAPTER 7

Beginning with the End in Mind

"Maybe the preoccupation with technological progress has overshadowed our concern with human progress."

Wynton Marsalis

The Humanist Revolution is not a siege of the castle, "off with their heads" kind of revolution. It's a silent one, a gradual coming together of millions of people to reset and reshape global governance and the underlying systems, using a shared definition of human progress and human understanding as our guiding lights. Ironically, the need for such a revolution has been driven by a series of other technology-driven upheavals that began with the Industrial Revolution (IR) in the 19th century. Since then, modern society has been slowly flooded by subsequent waves of technology and, along with them, profound life and work shifting revolutions, concluding most recently with, as it's known, the Fourth IR. In his talk at the World Economic Forum in 2016, Klaus Schwab, the founder and Executive Chairman, explained:

> *There are three reasons why today's transformations represent not merely a prolongation of the Third Industrial Revolution but rather the arrival of a Fourth and distinct one:* velocity, scope, and systems impact. The speed of current breakthroughs has no historical precedent. When compared with previous industrial revolutions, the Fourth is evolving at an exponential rather than a linear pace. Moreover, it is disrupting almost every industry in every country.

DOI: 10.1201/9781003089902-11

And the breadth and depth of these changes herald the transformation of entire systems of production, management, and governance.

Seven years later, Klaus's prediction of the transformation of entire systems of production has come true. On the other hand, his call for essential changes in our planet's management and governance have not.

In response, some have begun to theorize about a Fifth Industrial Revolution that may be both a contributor to and consequence of an overall Humanist Revolution that can force the needed transformation of how we manage and govern ourselves. In a recent article sponsored by the World Economic Forum, authors Pratik Gauri and Jim Van Eerden spell out the potential of this latest stage of the dance between humanity and technology:

In contrast to trends in the Fourth Revolution toward dehumanization, technology and innovation best practices are being bent back toward the service of humanity by the champions of the Fifth …(it) has the potential to initiate a new socio-economic era that closes the gaps between the "top" and the "bottom," creating infinite opportunities for humanity, and for a better planet.

The critical question unaddressed within their declaration is who exactly will bend technology back toward the service of humanity? Who exactly will forge and enforce the absolute policies to result in a collective and effective response to the existential threats at our door? Who will step forward to engineer and deliver the necessary resets, the essential recalibrations, and a whole global commitment to doing right by people and our planet before all else? It must be us.

Like all successful movements, peaceful or not, the Humanist Revolution must focus on two even more fundamental questions: What are we after exactly and how well designed and aligned are our current systems and behaviors to realize those outcomes? And if the answer is we don't really know what we seek, we need to. And if our systems and behaviors do not point to that newly defined outcome, they must be changed to do so. Because as pedantic as it sounds, our future, the future of humans and the planet we rely on lies in large part with the integrity of our intentions and systems design. It's not a new idea. Systems have long defined and guided our existence. Natural systems, social systems, manmade systems, systems of protocol, of technology, of production, of legality, of economics, and even religion are all carefully designed to achieve a variety of outcomes, from safeguarding our assets and yielding goods and services to satisfying our Maslovian needs for safety, control, and belonging.

Countries, companies, cities, families, and even you and I are implicitly system designers, integrators, and/or adherents, with many of our current systems having been edited, updated, and added to continuously over the decades and even centuries. Herbert A. Simon, the Nobel Prize-winning American economist, political scientist, and author of the seminal 1969 book, *The Sciences of the Artificial* called out the fundamental role that (systems) design plays in how our human world works when he wrote,

> Everyone designs who devises courses of action aimed at changing existing situations into preferred ones. The intellectual activity that produces material artifacts is no different fundamentally from the one that prescribes remedies for a sick patient or the one that devises a new sales plan for a company or a social welfare policy for a state.

A MATTER OF INTENTION

The capacity to design is the capacity to improve and because we are fallible beings, at times unintentionally impair. Our capacity to design successfully is predicated on our ability to first articulate the intention of the design clearly. Oddly enough, 200,000 years after the emergence of us, *Homo sapiens*, we still don't have a universal definition of human progress. The human existence system is beginning to fail because we don't understand and have never fully understood the point, the intention of the system. After the many centuries of innovations, of scientific breakthroughs, of gains in our philosophical and economic grounding, 21st century global society does not have a sufficiently comprehensive answer to the question "What are we after as a species?" And without a consensus definition of human progress, we cannot steward technology (to where?) and we cannot deal with the litany of technology's unintended consequences. Nor can we successfully combat the existential threats that seem to be popping up everywhere. And finally we cannot begin to re-design the underlying systems to be essential contributors to those desired outcomes. You simply can't make a system better if you don't agree on what better is.

In their 2018 research paper for the National Research Council of Italy titled "A Critique of Human Progress: a New Definition and Inconsistencies in Society", authors Mario Coccia and Matteo Bellitto underscore the complexity of the human progress question:

> Overall, then, the whole process of human progress is driven by the increasingly effective struggle of the human mind in its efforts to raise superior to the exigencies of the external world and attitude to satisfy human desires, solve problems and achieve/sustain power in a sustainable society. However, a comprehensive definition of human progress, at the intersection of vital elements of economics, sociology, psychology, anthropology, and perhaps biology, is a non-trivial exercise.

Non-trivial indeed, but still essential.

WHAT IS HUMAN PROGRESS?

Due to the lack of a shared definition of human progress and because of the current default position that it solely means economic (GDP) growth, we have allowed the technology train to get further and further away from our humanity, and what is actually good for us now and in the future. Absent the guiding governors of specific, measurable, and holistic human progress criteria, the "economic growth at any cost" technology train has become a reckless engine, an idea underscored by Coccia's and Bellitto's summary declaration:

> ...we believe that psychosocial factors of people in society have their vital weight in the debate on human progress. We assume that an advanced society must support mainly happiness, social wellbeing and sustainable environment, rather than a blind economic growth with consequential environmental, social and food security threats.

It has been a consistent theme in our all too human story. Our evolution, from primate to modern human, has lacked clear success criteria and collective, calculated intention. It has largely been a story of opportunism and adaptation not out of collective purpose but fueled by necessity, self-interest, and a great deal of capitalistic motivations. Along the way there has been little consideration for social, environmental, or humanistic consequences, except for reactionary periods including the 15th and 16th century Renaissance, the 17th and 18th century Enlightenment and the 19th century advance of humanism generally. The 20th century World Wars, coupled with the onslaught of unbridled technology, appears to have shifted the focus solely towards nation-state and individual self-interest.

As explored in Chapter 2, the undercurrent intention of most innovations since the beginning of humankind appears to have been a combination of more speed and greater longevity; both outcomes encouraged by a misbegotten sensibility that more of both was and always is better. As we now struggle in an age of too much and yet not enough, of too fast for our own biological microprocessors to handle, it's increasingly clear that these aspirations were both insufficient and even potentially counterproductive to the human progress effort. And, not surprisingly, they are decidedly poor gauges of the short-term viability and potential long-term consequences of the fire hose of new technologies and innovations spewing at us every day.

THE MISCALCULATION IN GDP

The idea of a country's Gross Domestic Product (GDP) as the central measure of societal progress first appeared in the United States in 1937. Partly in response to the Great Depression, it was based on a simple equation: if the total economic output of a country was growing, then the economic means of its individual citizens should grow too. More output meant more jobs and greater personal income. And more individual wealth meant greater collective progress. By 1944, the GDP had been adopted by scores of countries around the world as the primary measure of economic welfare, aka human progress. GDP remains the most generally accepted human progress measure to this day, even while more and more people in the developed world express dissatisfaction with their lives and our collective state. Over the 80 or so years of GDP "rule" there have been a growing number of voices challenging the basic math, proposing that a nation's success, its progress (and by extension the planet's), is tied to more than money. One of the most famous declarations came in 1968, from Robert F. Kennedy, then a Senator in the U.S. Congress, while speaking at a conference at the University of Kansas. His remarks, focused on the future of America, included this reference to the flaw in GDP as humanity's sole measure:

> …that Gross National Product counts air pollution and cigarette advertising, and ambulances to clear our highways of carnage. It

counts special locks for our doors and the jails for the people who break them. It counts the destruction of the redwood and the loss of our natural wonder in chaotic sprawl. It counts napalm and counts nuclear warheads and armored cars for the police to fight the riots in our cities. It counts Whitman's rifle and Speck's knife, and the television programs which glorify violence in order to sell toys to our children. Yet the gross national product does not allow for the health of our children, the quality of their education or the joy of their play. It does not include the beauty of our poetry or the strength of our marriages, the intelligence of our public debate or the integrity of our public officials. It measures neither our wit nor our courage, neither our wisdom nor our learning, neither our compassion nor our devotion to our country, it measures everything in short, except that which makes life worthwhile.

Twenty years earlier, the American journalist and essayist H.L. Mencken had presented a somewhat similar, albeit crisper, view. He wrote,

A country or population realizes progress when change results in any (or all) of the following: longer lives, reduced infant mortality, decrease in morbidity, increases in people's options, greater equality, more freedom or a reduction in fear of other people or of their own rulers.

Mencken's references to increased optionality, greater equality, more freedom, and the reduction of fear as component measures of human progress were prescient precursors to the progressive thinking that is unfolding today, thinking that is in a way the aspiration of the Humanist Revolution. A remarkable array of global organizations and countries are attempting to establish more comprehensive and human measures of human progress as both better gauges and as a rebuttal of sorts to the longstanding global fixation on GDP. Much of today's human progress measurement movement can be attributed to work done by Richard Easterlin, a professor of economics at the University of Southern California who, in 1974 exposed the fact that even with significant growth in personal incomes in America, happiness over the prior almost 30-year period had remained flat. His research explicitly proved that the direct correlation between human progress as life satisfaction and human progress as economic gain did not actually hold over time. With Easterlin's de-bunking of GDP as the sole measure of progress, happiness, and its figurative cousin "well-being" quickly emerged as alternate measures, spawning several new indices and gauges, including a longitudinal effort by the United Nations (UN) and its Development Programme (UNDP) first undertaken in the late 70s. As an introduction to a recent annual report, the UNDP wrote, "Instead of using growth in GDP as the sole measure of development, we ranked the world's countries by their human development: by whether people in each country have the freedom and opportunity to live the lives they value." The UNDP preface goes on to declare,

> …we are at an unprecedented moment in history, in which human activity has become a dominant force shaping the planet. These impacts interact with existing inequalities, threatening significant development reversals. Nothing short of a great transformation— in how we live, work and cooperate—is needed to change the path we are on.

The UNDP's call for a great transformation is a call for the Humanist Revolution, resulting in a 21st century renaissance that yields an entirely different approach to how we steward technology, our planet, how we engage with each other, and how we measure our progress on that path.

THE SUSTAINABLE DEVELOPMENT GOALS AS HUMAN PROGRESS

The UNDP's early work was a precursor to the 2015 issuance by the United Nations of its 17 Sustainable Development Goals (SDG), another more faceted effort to define and measure human progress. The result of a collaboration between the UN's 193 countries, the SDGs are an attempt to create a 15-year roadmap, a multilateral system with 17 clear measures (outcomes) that governments, NGOS, corporations, and grassroots organizations could implement to ensure prosperity for more people. The 17 SDGs, shown in Figure 7.1, range from the elimination of poverty and hunger to more affordable, clean energy.

The list passes common sense muster, reflecting the facets of life that appear to matter most. What the list does not capture is the question of whether the systems that underpin much of the work that needs to be done in each goal area, are effective systems, including the United Nations itself.

FIGURE 7.1   United Nations 17 sustainable development goals.

More pointedly, if the current systems of governance, capitalism, education, or health care are fundamentally broken, how could they possibly make progress towards the related goals?

The 17 SDGS also don't convey any sense of priority, as in which measures contribute to the achievement of other measures? Are they all independent propositions? Assuming linkage and a projected $45 trillion price tag and tight timetable, the OECD (in partnership with the think tank New America and Greenhouse, a social innovation group) undertook an extensive study to answer those questions. They interviewed 85 economists, sociologists, and experts from both the public and private sector to determine a prioritized ranking of the 17 goals. The outcome of their work is depicted in Figure 7.2.

The graph suggests that well-being is less important to the march or measure of human progress than most other attributes. But if we consider the prioritized list through the lens of cause and effect, of means versus ends, it's possible to recalibrate the ranking around the simple idea that well-being is more an outcome (intention) than a contributor to the system itself. It is that view which would hold up well-being as the primary output of all systems rather than just a metric of progress. Since we cannot directly act on well-being, i.e., dictate that it happens, the best that we can do is create circumstances that contribute to it. Interestingly, it's likely that technology has little, or certainly less, of a role here. While technology has been brilliant at fueling the functional part of Maslow's Hierarchy of Need,

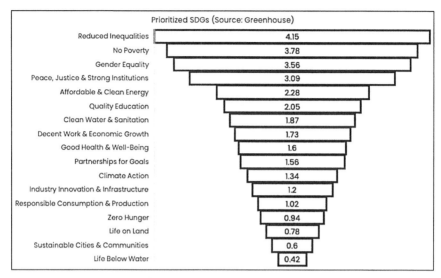

FIGURE 7.2   Sustainable development goal prioritization.

providing more people with shelter, safety, and some degree of belonging, it has failed mightily in its contributions to the intrinsic, hyper-personal matters of collective consciousness, self-esteem, and self-actualization, all major contributors to well-being, and to whether one is happy.

RE-DEFINING THE MEASURE

One of the most universally recognized happiness gauges is the World Happiness Report, an annual index first published in 2011 by a collaboration of research and academic institutions and now presented by the Sustainable Development Solutions Network for the UN. The study seeks to determine every country's relative happiness across six measures: GDP per capita or income, availability of social support from family and friends, healthy life expectancy, freedom to make life choices, generosity received over time, and perceptions of corruption, or more positively, trust in government. In 2023, according to the World Happiness Report the ten happiest countries were,

1. Finland
2. Denmark
3. Iceland
4. Israel

5. The Netherlands

6. Sweden

7. Norway

8. Switzerland

9. Luxembourg

10. New Zealand

The world's reigning number one superpower, the United States, came in #15, even though its per capita income far exceeded most of the other 179 countries ranked. In 2020 it ranked 10th for income but significantly underperformed on the other five measures. Surprisingly, the self-declared "land of the free" came in 61st on freedom to make life choices and 42nd on corruption. The mixed bag on U.S. performance affirms the now familiar adage: money cannot buy happiness. It also suggests that the system called the United States has a great deal of work to do.

Given the slow but growing interest in a more humanistic versus economistic view of human progress, is The World Happiness Report then the right set of measures? If more and more of the 8 billion people who make up the human species declare themselves to be happy, or at least are experiencing the attributes that correlate with happiness, aren't we making the right kind of progress? One perspective is that the six measures of relative happiness are certainly more dimensional and representative of our individual and collective interests than the blunt instrument called GDP, and therefore reflect material progress in measuring material progress. Another is that they don't go far enough to understand what humans seek relative to their current needs and future wants. That view brings the debate back to Maslow's Hierarchy of Needs and the idea that human progress should correlate with how much of a particular population thinks, feels, and can achieve life-stage correlated levels of need, from the physiological (food and shelter) to the self-actualized (freedom and achievement). As Richard Barrett, a pioneering researcher, and the Director of the Academy for the Advancement of Human Values, wrote in the preface of his book *Worldview Dynamics and the Well-being of Nations*:

> …I have discovered…that a person's sense of well-being is dependent on their ability to satisfy the needs of the stage of psychological development they are at… (which) is in turn dependent on the social environment in which the person lives—the values

and beliefs of their parents, the community in which they are raised, the school they attend, the workplace where they earn a living, the religion they adopt and the worldview of the nation where they live. The values and beliefs in each of these environments either support or hinder a person in getting their psychological needs met and thereby have a strong influence on their sense of well-being.

Barrett is indirectly calling out the system intention question and underscoring that the systems around us, the ways of our society as manifest in beliefs and behaviors, processes and policies, rules, and laws, contribute to the accomplishment of those intentions or not. He is also suggesting that one size may not fit all and that progress is relative depending on where a specific entity is in its journey. And also that the ultimate measure of progress is impacted as much by how people feel, what they believe, and how supported or not they are or perceive they are, as it is economic capacity. Like the 17 United Nations Sustainability Goals, there remains the seductive questions of primacy and sequence. Is there one indicator that is likely the first domino that *most* countries, cities and our planet should focus on? The ranking work done by the OECD on the Sustainability Goals established eliminating inequality as job one, following by eradicating poverty and reducing gender inequality. Implicit in these three top ranked goals is the idea of equal opportunity, or again, as H.L. Mencken called out, "…increases in people's options, greater equality, more freedom or a reduction in fear of other people or of their own rulers." The idea of opportunity, of economic and racial equality, of freedom and even the reduction of fear could be combined under the superset concept and measure known as "latitude." The dictionary definitions of latitude include freedom of action, freedom of choice, and freedom of thought. It is not hard to imagine a more progressed and humane world when most of its inhabitants believe they have, and do have, latitude, i.e., the latitude to both meet their needs and achieve their aspirations, however big or small, in whatever life stage they are in and world view they have.

As Jeff Leitner, a principal at Greenhouse, one of the OECD partners on the Sustainability Goal ranking work declared at the conclusion of the study:

> But for the moment, we conclude that development experts believe the best way to start working towards the SDGs is to expand people's opportunities to reach their own goals, with governments guaranteeing the rule of law, social stability and minimum conditions of life for all.

Leitner's declaration reinforces the idea that human latitude may be the linchpin to well-being, the superset measure of human progress. But that latitude must be enabled by policy-level structures and systems focused on Maslow's functional stages of need; i.e., systems that keep the productivity engine running while also enabling the realization of individual poten-tial (the higher stage needs) and collective contribution to the advance-ment of the whole. The traditional objection to such an approach lies with the question of whether most people have goals, seek opportunity to have impact, to realize their full potential. Are most of us, in Maslow's parlance, after "self-actualization"—do we really want to transform ourselves? In our answer may lie latitude's limitations as a universal measure. Because to further improve human progress requires a focused, systematic, and integrated approach that motivates and requires actions towards it. Even in a country as "developed" as the United States, it's possible that most of its citizens are content to not progress beyond making a little more money or having a few more possessions. To achieve our expanded definition of human progress will require all our systems, including our governing bodies, to play a central role in imbuing the idea of realizing individual potential and collective responsibility while functionally supporting those outcomes.

THE END IN MIND

To combat the myriad unintended consequences that technology has wrought, take on the long list of existential threats, and address the growing divisions and inequities in the world, it should be clear that we must begin with the end in mind. We must come together across countries and continents, across conflicts and disputes, involving organizations, commu-nities, and individuals, to form a shared view of what a better world for more people means. The UN's work on the 17 Sustainability Goals is a solid first step but it is insufficient in that it ignores causal relationships (means versus ends) and hesitates to prioritize the list. We need one goal that subsumes all other goals, with an emphasis on the truly human ones. The task then is to evolve the SDG thinking and force a collective global dis-cussion and agreement regarding how we will singularly define, measure, and march to our shared, desired destination: collective well-being and the latitude of choice it presumes. And with our agreement we can find ways to motivate participation by every potential contributor, from the Elon Musks of the world to your neighbor next door. Motivating versus demanding participation is key. And top to bottom participation is best accomplished by doing three motivating things:

(1) Establishing relevant and motivating shared intentions (the ends);
(2) Engaging people around the crafting of the means (the plan) as a story to be created together;
(3) Asking for their help in making the story come to life, to realize our intentions.

This three-part model of motivation and engagement applies to the Humanist Revolution. Our intention is collective human progress as defined by expanding well-being and latitude, where all people can realize their full potential, to meet their basic needs and realize their aspirations. The means begins with a new code of behavior, a set of humanity first principles we must all learn to live by. With that as foundation we can then take the next step, a thorough examination of all the systems we rely on and a determination of how well they reflect the new code and help us achieve our intention. And then to make the requisite changes to those systems to insure alignment.

For our plan, our newly designed "system of systems," to have global impact it will need to be adopted at every level of society. We must engage the help of governments, corporations, cities, communities, and everyday citizens. Universal participation and global conformance will only happen if most of us are motivated by the potential of the revolution and the humanist behaviors it calls for. It will only work if many more of us embrace our mutual reliance, our interdependence, and accept that borders are irrelevant, that untethered technology and capitalism may be our undoing, and that humankind and its survival is entirely dependent on our ability to re-define our intentions and the systems that will enable their achievement.

It's possible that the most dominant and troubling force on this planet is not human activity, but rather the lack of human activity related to technology and the need for its proactive, human-first stewardship. Our inability to mitigate the many unintended consequences of the technology train plowing through our lives is a direct consequence of not having a measure of what kind of lives we seek and not being willing to step forward to demand that innovations and systems contribute to that outcome. We need to replace the human activity that has created the mess we are in with the human activity that will get us out of it.

The necessary transformation of the way the world works and the way we are is in fact a human transformation. Realizing real human progress will not happen overnight, nor will the flip be switched everywhere in the same way. It will take decades if not centuries to catch up with the train. That intention-defining, catching up work cannot be carried out by one nation or global governing body. It will have to be fomented and driven by us, by the Humanist Revolutionaries, with a focus on the structural and systemic, mirroring the strengths of competitive capitalism but with an entirely different set of incentives and outcomes in mind. As this chapter serves as a first step towards re-defining our intentions, the chapters that follow are meant as an operating framework for the revolution, a description of the work to be done.

The point is the point.

CHAPTER 8

Establishing a New Paradigm of Rule

"Globalism is not mushy government idealism—far from it. It does not deny the existence or importance of government—at the local, state, national or international levels—or of intergovernmental diplomacy. But it insists that the great-power games, as deadly as they have been and could still be, must give way to planetary politics, in which human beings matter more than nationalities. Competition itself is fine and natural, but it needs to be competition to achieve a goal that benefits us all."

Anne-Marie Slaughter
CEO, New America

The current paradigm of rule can be characterized by a familiar idiom, "dog eat dog," the power-driven hierarchies that render a minority of winners and a majority of losers. The unrelenting advance of most technological innovation has been fueled not by what is truly good for the masses in the long-term, but often by what is good for the handful of technologists, investors, and corporations that bet on those innovations in the short-term. The rules of commerce, of governance, of societies, including morality and ethics, have been manipulated and modified by the "elite." The result is that our "ways" are increasingly losing touch with what really matters, i.e., how best to serve humans and to be human. The rules are a splotchy patchwork of ideals concerning human freedoms and dignity, layered over a checkered history of division, nationalization, paternalistic hierarchies, of incessantly warring tribes and states fighting to establish hard borders and distinct destinies. These "systems," many of which have been handed down from the

DOI: 10.1201/9781003089902-12

127

Enlightenment, are crudely grandfathered and disconnected propositions. They all appear to struggle to reset themselves in light of the new challenges and disruptive technologies. They are loosely based on legacy behaviors that are shockingly primal, focused on self-preservation, and the amassing of power, not the realization of progress for all. In fact, much of the progress we have made in learning to govern ourselves has been fueled not by moral human progress intentions but by channeling our instincts for self-preservation into collective progress structures—systems of law, markets, and democracy most prominent among them. We have effectively created an "invisible hand" that magically transmutes selfishness into some forms of economic progress. It's an observation first made by the 18th century Scottish political economist and moral philosopher Adam Smith:

> Every individual... neither intends to promote the public interest, nor knows how much he is promoting it... he intends only his own security; and by directing that industry in such a manner as its produce may be of the greatest value, he intends only his own gain, and he is in this, as in many other cases, led by an invisible hand to promote an end which was no part of his intention.

The invisible hand has delivered profound functional gains, but also some very human and environmental losses. And the capacity of that invisible hand to solve our collective problems is quickly diminishing. And this is why the paradigm of rule must be changed. We need a new paradigm to realize our more expansive definition of human progress, for future generations to not only survive but to thrive. We need to replace the invisible hand with visible hands, our hands.

As explored in Chapter 6, many of our legacy ruling structures and systems are already in an awkward, adolescent transition, with power models shifting and traditional institutions faltering and exposing their limitations in the face of challenges on a global scale. Thus, 20th century formations of multi-lateral governance structures are proving far too emasculated for the task while global corporations are emerging as a potent source of super-national power and influence. As more multi-national corporations insinuate their views into matters of public policy and global governance, the power of the existing governing bodies appears to be ebbing just as the number of existential threats and cross-border disruptions rise. It is a daunting picture well captured by Augusto Lopez-Claros, Arthur L. Dahl, and Maja Groff in their recent book *Global Governance and the Emergence of Global Institutions for the 21st Century*:

> In a globalized economy and society, improved global governance must play an important role at this crucial moment when change is increasingly urgent for environmental, social, and economic reasons. It is not easy to set priorities among the many challenges of today, as all are interrelated. Their complexity calls for new approaches suitable for dynamic, integrated systems evolving through constant innovation in technologies, forms of communications, patterns of organization and institutional frameworks. The challenge for mechanisms of governance at all scales of human organization is to accompany and steer these processes to ensure the common good, setting limits that prevent their being captured by the already rich and powerful for their own benefit, and ultimately ensuring a just society that guarantees the well-being of every person on the planet.

The authors are calling out three unsettling truths. First, the task is incredibly complex. Second, we have limited, if any, recent experience effectively steering processes (and systems) towards a common good at a national level let alone a global one. And third, we have little evidence of the ability to contribute to the common good without the elite taking their disproportionate share of the power and the wealth. A daunting picture indeed, but that is the substance of the Humanist Revolution: envisioning, innovating, trying, and improving the way the world works.

A MATTER OF MOTIVATION

The question is this: Do we have the desire and the capacity to change the ways we rule, operate, think, and behave to ensure a different outcome? I have long believed that there are only two viable motivators of behavior change, two motivators that cause humans to take wholly different paths at certain critical junctures: aspiration or desperation. Aspiration is a want state. Desperation is a need state. As you might imagine, the latter is far more effective than the former. Arguably, the desire for human progress, the achievement of the 17 United Nations Sustainability Goals, the democratic expansion of human latitude and opportunity, and the effective combatting of enemy number one, climate change, are all just aspirations, much like the moral matters of justice and equality. Alarmingly, the many existential threats we face, climate change, pandemics, cyberwarfare, negative birth rates, nuclear devastation, and more broadly, the unchecked, unintended consequences of technology, are not yet in the desperation category. Many governments, most leaders of corporations, cities, and communities, and most of us individually do not yet feel the need, the desperate pressure to combat our encroaching enemies with decisive action. We do not yet believe we must change our behavior and make wholly different decisions. In his 2020 book *The Precipice*, author and Oxford University professor of philosophy Toby Ord proposes that economics has a major role in dissuading governments from caring and doing more. His compelling argument is that existential risks are "undervalued by markets, nations and entire generations" because the economic return is distributed across borders and therefore the investments required by country for a "global public good" are disproportionate with each country's individual gains. Once again, self-interest at a nation state level is in the way. He explains, "Since such a large proportion of the benefits spill out to other countries, each nation is tempted to free-ride on the efforts of others, and some of the work that would benefit us all won't get done." He goes on,

> This means management of existential risks is best done at the global level. But the absence of effective global institutions for doing so makes it extremely difficult, slowing the world's reaction time and increasing the chance that hold-out countries derail the entire process.

Ord confirms a central belief: the greatest threat to humankind remains the inability of humankind (individually and collectively) and particularly our leaders to assertively collaborate to combat our existential foes. The near-term enemy appears to be us.

REAL OR UNREAL

We cannot even agree on whether the threats are real. As but one example, for all the climate science and research, for all the real-life evidence of entire cities and regions being decimated by fire, water, drought, and disease, there remains a large percentage of the general global populace and its many types of leaders who continue to deny that climate change is a cause for concern and in some cases that it even exists. According to a March 2021 analysis conducted by the Center for American Progress, an independent non-partisan policy institute in the United States, "there are 139 elected officials in the 117th Congress, including 109 representatives and 30 senators, who refuse to acknowledge the scientific evidence of human-caused climate change." Three years later the statistics have not changed much.

The climate change deniers around the world are joined by millions of COVID-19 and vaccination deniers, not to mention the many who now distrust any policy promulgated by the scientific elite. The capacity to deny is both a universally human one and a stance fomented by technology. As technology has muddied the capacity to separate truth from falsehood, science from silliness, it has given the masses the license to deny pretty much anything that would either require them to behave differently or infringe on their personal freedoms. And even among those who believe that the threats are real, their beliefs and words are often not supported by their actions and behaviors. As discussed, it's in part a biological phenomenon, the "fight or flight" amygdala kicking in to overrule our frontal lobe to arrive at the decision that doing nothing is the best course of action. But it's also psychological, as evidenced in a 2009 study led by the American Psychological Association (APA) on psychology and human response to climate change. The research determined that our unwillingness to do anything differently because of climate-based threat is fueled by five psychological forces:

- **Uncertainty**—Research has shown that uncertainty over climate change reduces the frequency of "green" behavior. The flood waters are not yet at our door.

- **Mistrust**—Evidence shows that most people don't believe the risk messages of scientists or government officials.

- **Denial**—A substantial minority of people believe climate change is not occurring or that human activity has little or nothing to do with it, according to various polls.

- **Undervaluing Risks**—A study of more than 3,000 people in 18 countries showed that many people believe environmental conditions will worsen in 25 years. While this may be true, this thinking could lead people to believe that changes can be made later. We tend not to think beyond the months ahead.

- **Lack of Control**—People believe their actions would be too small to make a difference and choose to do nothing.

This last force listed, "Lack of Control" is perhaps the most interesting facet of our reluctance to act, and the most troubling. It suggests that most humans believe that the threats (whatever they might be) are so great, so overwhelming in their magnitude, that personal effort is of little use. In a recent interview with *The New York Times*, the author and chief scientist for *The Nature Conservancy* Katherine Hayhoe, captures the personal paralysis of many regarding combatting climate change:

> We're so individualistic that it affects our perspective on climate solutions…you could feel like you're doing everything possible to cut your personal footprint, and that would be a tiny fraction of the solution. When we realize that as individuals we can't fix it, that's when despair comes in. And if we don't realize that we have a shadow, not just a footprint, we feel like there's nothing that we can do because all we see is companies squabbling and avoiding action and the government in deadlock. I mean I get discouraged…

OUR GLOBAL GOVERNING BODIES

Our unwillingness to sacrifice (status, power, prestige, comforts, conveniences) to advance the cause of our planet and species is at the root of our reluctance to re-think, re-architect, and materially change how we steward the world, technology, environment, and humankind itself. The current major global governing bodies, beginning with the United Nations and including the World Health Organization, the World Bank, and the World Trade Organization all operate not as implementers but as influencers. They serve as the lowest common denominator risk mitigators versus the needed visionary motivators, enablers, and enforcers of the policies required to steadily achieve collective human progress. It's a limited purpose that has changed little since the very beginning of the formation of the global guidance system in 1947 with the founding of the United Nations (UN). Dag Hammarskjöld, the UN's second secretary general made the point when he said, "(The United Nations) was created not to lead mankind to heaven but to save humanity from hell."

Borrowing Hammarskjöld's vernacular, the intention of the Humanist Revolution is a figurative form of heaven: the realization of greater well-being for all. Hell is an unacceptable target. The current dilemma with the U.N.'s longstanding, limited, and largely reactive global governance position is that the intensity and multiplication of existential threats now require persistent proactivity, risk amenable innovations, and the capacity to enforce unpopular policies worldwide for the sake of our species. The lackluster performance of the World Health Organization (WHO) during the COVID-19 pandemic is a great example of a global governor standing on its passive, largely impotent heels instead of its declarative, empowered toes. That impotency indirectly contributing to the deaths of more than 5 million people worldwide. A May 2021 article by the BBC titled "Covid: Serious Failures in WHO and Global Response", revealed the results of a WHO sponsored panel's report on the failures of the organization itself and global governance in general throughout the pandemic, beginning with the declaration that the "the combined response…was a toxic cocktail." The article and the report both explicitly and implicitly capture the underlying flaws in the global system, flaws that need to be corrected if we are ever to win the battles, let alone the wars. The BBC journalists wrote:

> When countries should have been preparing their healthcare systems for an influx of Covid patients, much of the world descended into a "winner takes all" scramble for protective equipment and medicines, the report said. To prevent another catastrophic pandemic, the report suggests key reforms:
>
> - A new global threats council should be created with the power to hold countries accountable
> - There should be a disease surveillance system to publish information without the approval of countries concerned
> - Vaccines must be classed as public goods and there should be a pandemic financing facility
> - There was an immediate request for the wealthy G7 nations to commit $1.9bn (£1.3bn) to the WHO's Covax programme providing vaccine support to low-income countries

Panel co-chair and former New Zealand Prime Minister Helen Clark concluded by declaring that it was "critical to have an empowered WHO."

Some have proposed that the COVID-19 pandemic was a relatively benign threat in comparison to climate change and other invisible enemies, serving as a sort of test run of our global response system. It's a test that our current paradigm of rule failed. The WHO COVID-19 panel's prescription for how to better prepare for and respond to the next pandemic is, arguably, a viable global governance framework for how to better plan for and combat most if not all the multi-lateral, existential threats we face. Consider the following:

It presupposes a shift in power from individual nations to the planet and the establishment of a new paradigm of rule designed for the good of the whole versus one that perpetuates the ability of the nations at the economic top to retain their position. The key words in the panel's report, "empowerment," "accountability," and "without the approval" all presume (hoping against hope) that all countries will be willing to sacrifice, to honor their collective responsibility, and to adhere to a different paradigm of rule. The "winner takes all scramble" initially referenced as undermining a more coordinated, effective pandemic response is a central philosophy embedded in the current paradigm of rule, a philosophy further codified by the adoption of capitalism as the central driver of most developed economies. The winner takes all is baked into the source code of capitalist economies. And even when nation states throughout history have attempted to install more egalitarian economic models and forms of governance, including socialism and communism, they seemingly have always devolved into a bastardized, but largely the same, winner takes all (as in dictator, party leader, or ruling class) scramble and outcome. While pundits argue the nuances, China, an avowed communist country ruled by one party, is decidedly fueled by self-interest, market demand, and an unquenchable thirst for amassing power. It wants to take all. The universal quest for power, explored in Chapter 6, is a major barrier to a more effective paradigm of rule, a stubborn human condition provocatively and darkly conveyed in one of the 19th century German philosopher Friedrich Nietzsche's works:

> What is good? Everything that heightens the feeling of power in man, the will to power, power itself.

> What is bad? Everything that is born of weakness.

> What is happiness? The feeling that power is growing, that resistance is overcome.

> Not contentedness but more power; not peace but war; not virtue but fitness (Renaissance virtue, virtù, virtue that is moraline-free).

GLOBALISM, NATIONALISM, AND SACRIFICE

As noted, power and its motivator, self-interest, are the frequent companions to our unwillingness to sacrifice, to share, to give up so that others can get more. And it may be our species' most problematic behavior. In contemplating the possibility of a no longer emasculated global governance system it's almost impossible to imagine any nation state stepping forward and making the choice to give up its power and right to choose, to defer to the greater good. The prospects of re-inventing the global governance system to give it more teeth and, along with that, more power to impact our future and fate are poor. Given that, should we consider the opposite tack? A full-on embrace of hyper-nationalism and the idea of every nation state holing up, walling up, and doing just what serves them best? History and economic example suggest that given our human tendencies for self-protection, nationalism is the only form of governance that works over the long haul. Its inherent divisiveness and some would suggest supremacist tendencies are counterbalanced by claims that it serves as the most efficient means of a broad-based sharing of wealth and collective economic gain. The nation that sticks together is presumed to prosper together (never mind that our concept of "nation" is riddled with ethnic and racial prejudices and socio-economic stratification). The nuances of nationalism versus liberalism are not the subject of this book, and their merits or demerits should be debated elsewhere. What does seem clear is that the presentation of nationalism, and hyper-nationalism, as optional paradigms of rule and the only way forward is made with little consideration of the growing global specters of threat, the rapidly emerging, potentially unvanquishable enemies that know no borders and really do represent weapons of mass destruction. How can we embrace building our walls higher when our enemies are both shared and are already inside the gates? And equally troubling, as we build the walls higher, aren't we effectively imprisoning ourselves?

Paradoxically, within the walled confines of a nation state, hyper-nationalism and its frequent companion, autocratic rule, may be the most effective way for a *specific* country to combat many of the existential threats it faces in the very short term. China's initial response to COVID as one example, was entirely dictatorial and for democratic nations, extreme. And yet, it was remarkably effective in the short-term. Seventy-six days after the coronavirus was first detected in Wuhan, its 11 million inhabitants came out of a full press quarantine and returned to full production. Thanks to the shared sacrifice of some individual freedoms, China's GDP grew 7%

the following quarter. Compare that to America's performance during the pandemic and its federated model's dysfunctional inability to get legislative leaders, the 50 states, and many of their citizens to agree to any standard pandemic response protocols. As reported on January 2022, the United States vaccination rate was 62.6%. China's was 82.5%. The qualifier is that China's vaccines were far less effective, resulting in its short-term success turning into a longer-term problem.

The inference is that democratic, globalist rule is likely to underperform autocratic, nationalistic rule when it comes to both making and enforcing the hard decisions that are required to combat the existential threats we face. The Chinese government's recent decision to require video gaming companies to limit gaming time for youth under the age of 18 to three hours per week, from 8pm to 9pm on Friday, Saturday, and Sunday, (having once cited video gaming's function as "spiritual opium") both clearly portrays one of technology's unintended consequences as a threat and the challenge democracies will likely face in attempting to combat it and the other existential enemies we face. An interview with an American parent with young children in a September 2021 Reuter's article on the topic perfectly captures the almost inevitable collision between freedom of choice and what might be a very effective way to solve modern society's troubling screen-time problem:

> "Oh, that's an idea (China's three-hour video gaming law)," Duttweiler (parent), who works in public relations at a nonprofit, recalls thinking. "My American gut instinct: This is sort of an infringement on rights and you don't get to tell us what to do inside of our own homes. On the other hand, it's not particularly good for kids to play as much as even my own children play. And I do think it would be a lot easier to turn it off if it wasn't just arguing with Mommy, but actually saying 'Well, the police said so.'"

The limits of more autocratic, nationalistic approaches to addressing global existential threats are obvious. One country's solution does not solve the rest of humankind's problems. And even if humankind steps forward to accept that extreme challenges may sometimes warrant extreme actions, including rigid policies that squash individual freedoms but do get to collective, effective solutions, it's hard to imagine a full embrace of that proposition by the 195 nations and the eight billion people that make up today's world. The nature of us simply does not do well under the iron fist of others, for long, as noted by University of Michigan Professor Ronald Inglehart, the author of *The Silent Revolution: Changing Values and Political Styles Among Western Publics and Culture Shift in Advanced Industrial Society*. Inglehart calls out some supporting trends emerging over the last 50 years:

> A major component of the postmodern shift is a… declining emphasis on all kinds of authority. Deference to authority has high costs… under conditions of insecurity, however, people are more than willing to do so. Under threat of invasion, internal disorder, or economic collapse, people eagerly seek strong authority figures who can protect them. Conversely, conditions of prosperity and security are conducive to tolerance of diversity in general and democracy in particular. This helps explain a long-established finding: rich societies are much likelier to be democratic than poor ones. One contributing factor is that the authoritarian reaction is strongest under conditions of insecurity.

Inglehart's words point to a double bind situation. If the threats we face create greater insecurity, more of us will gravitate towards autocratic rule, a growing trend well underway, and autocratic rule by and large is hyper-nationalistic and not concerned with the global good. If the threats we face do not result in increased insecurity, then fewer of us will be willing to step forward to join the revolution, to attempt to re-make the paradigm of rule and accept the threat-mitigating policies that paradigm must produce. The answer to the double bind is more of us understanding that relying on an autocrat and/or standing still is not an option. Inglehart's proposed silent revolution is our Humanist Revolution.

THE PATH TOWARDS A DIFFERENT PARADIGM OF RULE

So, what is the way forward? What are the viable, practical actions to a different governing means, a different set of ruling structures and operating systems that will be effective at both combatting the many global threats while contributing to the steady march of truly human progress? If an empowered version of the current global governance institutions is an unlikely candidate, and the growing nationalistic autocracies have no interest in playing nicely, what then? Absent the fear fueled motivation of desperation, what will draw more of us together to forge a new paradigm of rule, one that puts humanity first and technology second, one that serves all, not just the few. What is the necessary model of governance capable of thwarting and even vanquishing the enemies at our door to ensure the future of humankind, at least for the next 100 years?

The only logical, plausible answer is us, the Humanist Revolutionaries. The motivation and effort to establish a new paradigm of rule will likely never come from the current rulers, the calcified governments, or emasculated global agencies. The systems are entrenched, and the divisions are insurmountable. The tenures of most political leaders are tenuous and often short, their self-interested quest for power remains paramount. The current paradigm of rule, at every level and in every form, is fundamentally incapable of remaking itself to respond to the threats we face and will only be willing to change when it's all too late. The new paradigm will be born of different leadership, from the rule not of the few but the many. We need to step forward, and in doing so focus our broad-based developmental energy on a radical re-vamp of policies, foundation systems, core practices, and human behaviors that begin not at the global or national government level but at a corporate, community, and individual one.

The revolution is human and silent because the intention and method will be focused on humanity and come from the ground up, from our human-first actions not politicized words. Such actions must first be exhibited by the leaders of the top 100 for-profit and not-for-profit organizations in the world that are the biggest creators, carriers, and enablers of economic prosperity and societal health. (As of 2022, global GDP was $95 trillion. The U.S. represents 20% of that number. The top 100 for-profit companies' market value represented over 30% of it). In addition to being the biggest economic drivers, these companies and their leaders have the unique capacity to motivate, guide, and support millions of people, as customers, employees, partners, suppliers, and shareholders. As revolutionaries, these leaders will have to accept their responsibility to move beyond the balance sheet and their business interests to help establish and deliver on a global agenda that results in a multi-lateral response to the threats we face, a unilateral human-first stewardship of technology, and shared agreement on the definition of human progress and a longitudinal commitment to investing in that desired outcome.

Global corporations are beginning to explicitly, if carefully, step into this global leadership role, indirectly contributing to the formation of a new paradigm of rule. In 2019, the Business Roundtable, an association of the chief executive officers of America's leading companies announced a new Statement of Purpose of a Corporation, signed by 181 of its members. The statement acknowledged the essential role of the corporation and its leaders in contributing to the long-term health and well-being of the U.S. economy and society at large. In response to the statement's issuance, Darren Walker, the President of the Ford Foundation, underscored its expansive scope:

> This is tremendous news because it is more critical than ever that businesses in the 21st century are focused on generating long-term value for all stakeholders and addressing the challenges we face, which will result in shared prosperity and sustainability for both business and society.

The obvious flaw in the Business Roundtable's expansion of its purpose is that statement's context remains nationalistic. Its promise is to America, while our biggest problems are global. And it presumes that all shareholders will accept that contribution to collective human progress is as important as contribution to the bottom line. (Remember the author of *The Precipice*, Toby Ord's explanation that the economics of the global public good simply aren't sufficiently motivating.) We are back to self-interest as the vexing condition and greatest hurdle to a more human and sustainable future. With that, some titans of the investment world, including Larry Fink, the CEO of Blackrock, the world's largest asset manager, are seemingly supporting the Humanist Revolution by calling on corporate leaders to lead purpose-led social change given "the failure of government to provide lasting solutions" while declaring that,

> Purpose is not the sole pursuit of profits but the animating force for achieving them. Profits are in no way inconsistent with purpose—in fact, profits and purpose are inextricably linked.

In a talk and interview as part of the *Computers, Privacy & Data Protection Conference* in February 2021 Tim Cook, the CEO of Apple, appeared as another Humanist Revolutionary, calling for both a more coordinated global response to cyber-terrorism and data abuses by suggesting that the European Union's General Data Protection Regulation" "...should be the law around the world" and declaring that we must put our humanity before technology and commercial gain:

> As I've said before, if we accept as normal and unavoidable that everything in our lives can be aggregated and sold, we lose so

much more than data, we lose the freedom to be human. And yet, this is a hopeful new season, a time of thoughtfulness and reform.

As Humanist Revolutionaries the private sector leaders around the world will need to support their purpose-filled words with more explicit, collaborative, and even sacrificial actions. As is true of all systemic change initiatives, it requires the creation of a global plan and timetable for how the for-profit sector intends to marshal the combined energy and influence of their companies to force global policy determination and implementation that results in material progress both against the myriad threats and towards a shared definition of human progress. Instead of spending billions of dollars lobbying in their own self-interest, their cause must become our cause and their plan must become our plan. This is a unique moment in time, when the paragons of global capitalism will have to put down their winner-takes-all swords and be willing to forego bottom line gain for collective benefit while pushing out the free riders. It's likely a question of who goes first, and how much pressure is being applied from the constituents they serve. That's where the rest of us revolutionaries come in.

With the revolutionary leaders of the Global 100 and the world's most powerful companies taking on more of the responsibility for human progress determination and global governance, so too must the NGO 100, the largest not-for-profits in the world, re-consider their role and relationship with and within the existing paradigm. In his provocative white paper titled "When It Comes to Global Governance, Should NGOs Be Inside or Outside the Tent?" author Mark Butcher calls out the inherent dilemma for NGOs who seek to operate within the current paradigm of rule or outside of it. Butcher acknowledges that for each NGO there are a multitude of tradeoffs and situation-specific variables that make the binary choice a false one while also suggesting that the trend line is towards a "process of private governance in pursuit of social change and economic justice." Like the Global 100, the NGO 100 should increasingly accept that hiding inside the current global governance tent while the enemy rampages outside, is no longer the most effective path. They too can become Humanist Revolutionaries and actively support the efforts of their for-profit colleagues.

Even with that emerging clarity and hope, there remain tough realities and challenges to how the Global 100 and the NGO 100 can do their jobs while taking on the job of helping to govern and guide the world. A January 2020 *Harvard Business Review* article titled "How Global Leaders Should Think About Solving Our Biggest Problems" by authors Mark Kramer, Marc Pfitzer, and Helge Mahne suggested that the two bedfellows of organization performance and societal impact are often decidedly not, "Instead, these initiatives collapse under their own weight as partners become discouraged by the lack of meaningful progress for society or economic benefit to the company." They go on to propose that the key to achieving bi-modal success is directly tied to how targeted the initiatives are to specific regions that also align with their organizational interests. Their simple summation: "Local solutions are the essential (means) to tackling global problems." The authors' point is that it's only by directly tying a corporation's market-specific commercial opportunities with the social structural challenges in those markets that their bi-modal efforts can deliver both performance and human progress.

The construct of global players driving local solutions points directly to the need to align the leadership role and reach of the biggest companies and NGOs in the world to the insights and energy of the tens of thousands of local community and business leaders around the world. That linkage is key. These on-the-ground Humanist Revolutionaries have the capacity to engage their citizenry, neighbors, employees, and friends in context-relevant and actionable ways. They have a unique ability to establish motivations, collectively demand policy reformation, exhibit new codes of conduct, and encourage collaboration towards a common good. Imagine a world where most governors, mayors, civic leaders, church leaders, school officials, and local business leaders agree on a shared and expansive definition of human progress. And then work towards its realization at every level of society while collaborating against the rising threats and unwinding the many unintended consequences technology has wrought. Imagine an alignment of intention and action among millions of leaders around the world as a means of radically changing the trajectory of us and our planet.

In affirming support of this growing groundswell is the emergence of several new not-for-profit organizations focused on developing the next generation of community and business leaders to play a more active role in the stewardship of the world, many focused on contributing to progress regarding the UN's 17 Sustainable Development Goals. One such organization is the Global Leadership Challenge (GLC), a joint initiative between the University of Oxford and the St. Gallen Symposium. GLC's declared vision well captures the underlying task of imbuing tomorrow's leaders with the sense and sensibilities they need to help govern the world:

> The Global Leadership Challenge aims to help emerging leaders to grow in the wisdom and character required for responsible leadership that makes a difference in the world—leadership that doesn't simply seek to fulfil personal ambition but furthers societies' sustainable development.

It is a vision statement that could be applied to the Humanist Revolution itself, since it points not just to our intention but the humanistic means to get there: the elevation of the common good above self-interest, the deepening of our wisdom, morality, and character and the imperative of emerging leaders from every facet of society to make the revolution happen. Our leadership must force a re-focusing and re-defining of the structures of global governance *and* a wholesale examination and re-engineering of the systems that underpin them. These are the foundation systems that make our world run, including the system of our own human behaviors and in particular, our relationship with technology and each other. This is the next step in the work we will do, in crystallizing and activating the Humanist Revolution.

A different end is dependent on an entirely different means.

CHAPTER 9

Re-engineering Our Systems

"When the rate of change inside an institution becomes slower than the rate of change outside, the end is in sight. The only question is when."

Jack Welch

As shared, our woefully one-dimensional definition of progress as economic growth and the grandfathered nature of most systems and their legacy ways are resulting in a slow, subtle, but ultimately debilitating set of system failures. In this now exponentially accelerating moment, with our planet under siege, our systems, once essential assets, appear on their way to becoming sclerotic liabilities. And because there are few alternatives, these flawed, disconnected, and siloed structures continue to operate as they are, effectively underperforming the real needs of our species. Consider for a moment how some of our core systems are currently configured, how well they contribute or do not contribute to a broader, more expansive definition of human progress like collective well-being and what they stubbornly hold onto as their measures of success. Health care systems put far more emphasis on remediation versus preventative, whole person improvement. The financial system holds up the performance of the stock indexes when most people don't own stock. The technology sector celebrates 5G as progress when 100 million Americans and much of the developing world either do not have access to broadband Internet or simply can't afford it.

All our core systems, including education, health care, finance, governance, and social systems are metaphorical ivory towers at a time

DOI: 10.1201/9781003089902-13

when interdependence and interconnectivity are required costs of doing business. America's political system is also an ivory tower, or more accurately, a collection of ivory towers. The legislative branch of the U.S. Federal Government is a tower that houses the two parties' separate towers: the Democrats in theirs, the Republicans in theirs. The twin towers and their constant partisan vitriol underscore the lack of a shared set of aspirations, of shared intentions, for America's future and that of the rest of the world. Partisanship is becoming more and more extreme in America politics and across the world in part because it is virtually impossible for any two groups or individuals to agree on a means if they don't first agree on the ends. A shared agreement on the definition of human progress is the essential first step to take the ivory towers of modern society down and replace them with a structure that is wholly different.

THE RISE OF UTILITARIAN INDIVIDUALISM

Of all the systems that set the tone for the other systems and their misguided, legacy intentions, the capitalist system may be the most influential and problematic. As defined by the British economics professor Tejvan Pettinger:

> Capitalism is an economic system based on free markets and limited government intervention. Proponents argue that capitalism is the most efficient economic system, enabling improved living standards. However, despite its ubiquity, many economists (increasingly) criticise aspects of capitalism and point out its many flaws and problems. In short, capitalism can cause—inequality, market failure, damage to the environment, short-termism, excess materialism and boom and bust economic cycles.

Implicit in Pettinger's description of capitalism is a collision between the idea of improving living standards (a common if incomplete corollary with human progress) and the desire for attributes like greater equality, less market upheaval, less materialism, and more sustainable environments. If both sets of "needs and wants" are critical, then our current system is failing mightily. If not, then increasing Gross Domestic Product, the current generally accepted measure of nation-state performance is the only measure of human progress ever needed, and we're doing just great.

With capitalism and its incumbent mentalities at the center of the modern world's operating equation, it's understandable that over the last century the other systems have become largely oriented towards enabling functional and individual economic gain, e.g., getting the patient back to work, educating the young to get a job, and putting the policies in place that support free markets. Few vocations, organizations, and systems in the developed world have been able to escape the constant call to make money, and ideally make more money. Bellah, in *The Good Society*, calls the combined systems outcome of the last 50 years "utilitarian individualism." His proposition is that the combination of capitalism and the full-on embrace of science and technology have denuded our core systems of their originally humanistic, common good intention, and responsibility. He argues that one of the greatest unintended consequences of technology and capitalism's unrelenting push is a system that is singularly focused on creating humans who create wealth for themselves, not humans who create lives of meaning and purpose for self and others.

In their 2016 *Harvard Business Review* article titled "The Problem with Legacy Ecosystems" authors Maxwell Wessel, Aaron Levie, and Robert Siegel underscore the challenge of transforming longstanding systems to meet new demands and complexities fomented and exposed by technology. They wrote,

> All companies' internal systems—their metrics, resource allocation processes, incentives, approaches to recruitment and promotion, and investment strategies—are set up to support their existing business models. These systems are generally well established and extremely difficult to change, and they often conflict with the needs of digital business models.

The company context the authors reference can easily be replaced with that of a city, country, or planet. The point is the same. The systems of the world are a motley collection of longstanding capacities, processes, agreements, metrics, and incentives. And they are almost blindly bound to legacy operating models and the accompanying functional and behavioral norms established over years. The rigidity of these systems and their overseers' resistance to change may have at some point been a strength but is now their greatest weakness. As more of us become clearer regarding an expanded definition of human progress, as our needs and concerns change, as the threats to civilization encroach, and more disruptive technologies are presented, the limits and issues of the legacy systems are being painfully exposed. It is another gap between technology and us, a picture confirmed by the author and entrepreneur Azeem Azhar in his recent book *Exponential*:

> Put together these two forces—the inherent difficulty of making predictions in the exponential age, and the inherent slowness of institutional change—and you have the makings of the exponential gap. As technology takes off, our businesses, governments and social norms remain almost static. Our society cannot keep up.

Our society and its defining systems are not keeping up and so the gap grows wider, and the fallout expands. The only way to close the gap and to reduce that fallout is to push for the decalcification and re-building of our legacy systems or more extremely, replacing them. This too is the work of the Humanist Revolution.

THE SEVEN I'S: A FRAMEWORK FOR RE-BUILDING OR REPLACING OUR SYSTEMS

Whether the revolution calls for re-engineering or wholesale replacing our core systems, there is a framework to help the designers and engineers get there. Developed during my time at Harvard, "The Seven I's" is a codification of the seven essential criteria that any modern-day system must meet to ensure its ongoing relevance, adoptability, and impact. An explanation of each "I" follows.

THE FIRST I: INTENTIONALITY

As explored in Chapter 7, "Beginning with the End in Mind", intentionality is a core requirement of any successful system, including the system of systems, human existence. Therefore, the first "I" in the framework is the essential step of re-setting each core system's intentions, with direct consideration of our broader human progress intentions. We simply cannot achieve the latter without re-focusing the purposes of the former. Clearer, more relevant, and human-centric intentionality will immediately force a reckoning of the foundation systems that we rely on: economic, political, educational, financial, and health care, and the way they do or don't deliver against those outcomes. Again, Robert Bellah and his co-authors of *The Good Society* underscore both the systems' shared problem and the solution:

To imagine them as autonomous systems operating according to their own mysterious internal logic, to be fine-tuned only by experts, is to opt for some kind of modern Gnosticism that sees the world as controlled by the power of darkness and encourages us to look only to our private survival…(we need a) public philosophy less trapped in cliches of rugged individualism and more open to invigorating, fulfilling sense of social responsibility…It is certainly not enough simply to implore our fellow citizens to "get involved." We must create institutions (systems) that will enable such participation to occur, encourage it, and make it fulfilling as well as demanding.

Theirs is a call for *all* core systems to expand their intentions to include an explicit contribution to the common good while motivating us as individuals to care about and contribute to that good. This thread of human consideration and intent was once a founding principle of most of the systems modern society built. It is an essential thread lost in the exhaust of the technology train roaring ahead, effectively replacing the good of the whole with the economic empowerment of the individual. The Humanist Revolution must work to restore this thread and weave it through every aspect of modern society, beginning with our systems.

Imagine the consequences of re-setting the intention of the entire education system to be focused on developing real-world, whole person competencies, mindsets, and behaviors, for life. The entire system would need to be completely re-configured from its current inside-out rigid set of traditions, standards, protocols, and incentives to outside-in, dynamic, connected, and real-world relevant. Consider the impact of a global political system focused not on short-term power mongering but on long-term shared human progress. Or financial systems built not just for feeding the capitalist beast but actively working towards financial inclusion and a fairer distribution of wealth. How about an American health care system designed not just to remediate disease at scale but to focus on the health of the whole person in a preventative, personalized, and lifelong fashion? Marianne Williamson, the author, political activist, and a 2024 candidate for the presidency of the United States, described the current health care system in the United States more vividly:

The biggest problem with America's health care system is that it is not a health care system so much as a sickness care system. It reflects an outdated perspective on health and healing, in which far too little attention is given to the actual cultivation of health and prevention of disease.

The shift to a genuine health care system would involve attention to environmental, agricultural, chemical and nutritional factors which America's current corporate-dominated system of governance would presumably resist. Yet if America is to deal with our serious issues involving chronic disease and obesity, we must look deeply at the causes of disease and not simply their treatment.

Williamson's statement exposes three overlapping issues: the system's flawed intentions, an unwillingness to look at the causal issues holistically and humanistically, and an entrenched form of capitalism first rule that values the here and now far more than the future. These are issues that plague all core societal systems.

As we contemplate the possibilities of a health care system focused on entirely different outcomes (and inputs) we should also consider the consequences of a global economic system that is focused not on GDP but on well-being and economic sustainability in the most expansive sense, beginning with providing all humans with a basic level of income. Universal basic income may be the most powerful idea within the set of re-calibrated system intentions, even though it has been debated for centuries. And while it has often been associated with the perceived downsides of an expanding welfare state, more and more private and public sector leaders are accepting that it could serve as the central driver of human progress by giving all people the latitude to be absolved of the abject, constant fear for survival. Exempt from having to desperately satisfy Maslow's first level needs of food, water, and shelter, the roughly 700 million adults and children who live in extreme poverty, surviving on less than $1.90 a day, would be able to focus on tending to their lives and realizing their potential. In his 1795 letter prepared for the government of the French Republic during the French Revolution, the political activist and philosopher Thomas Paine presented the concept of universal basic income as a matter of essential justice and human right. He wrote,

It is not charity but a right, not bounty but justice, that I am pleading for. The present state of civilization is as odious as it is unjust. It is absolutely the opposite of what it should be, and it is necessary that a revolution should be made in it. The contrast of affluence and wretchedness continually meeting and offending the eye, is like dead and living bodies chained together.

As uncomfortable a metaphor as Paine's may be and although it was written over 200 years ago, it still bears truth. To push for a broader definition of human progress, for greater human well-being and latitude, and to drive towards that outcome in every aspect of how things we work, is fundamentally, most humanly, a matter of justice. And it is justice that is at the core of the Humanist Revolution and perhaps has long been the motivator behind most revolutions. In his speech titled "Remaining Awake Through a Great Revolution" given at Washington D.C.'s National Cathedral on March 31, 1968, the civil rights activist Martin Luther King Jr. (MLK) declared,

> We shall overcome because the arc of the moral universe is long, but it bends toward justice.

Human progress must be just and justice must accommodate every measure of that progress. The "We" MLK calls out is us. It is the Humanist Revolutionaries that are responsible for bending the arc. Not governments, not global organizations, us.

THE SECOND I: INNOVATION AS INFINITE ADAPTATION

As we work to bend the arc and push for systems to be re-designed to meet more just, more human intentions, we will also challenge our systems to manifest the second "I" criteria, innovation as infinite adaptation. Occasional innovation is insufficient. Our systems must learn to innovate constantly, to become infinitely adaptive. In an ever-accelerating, constantly changing world, sustained relevance demands constant re-invention. The shelf lives of systems value, systems efficiency, and systems impact are shrinking. The leaders of these systems, their stewards and stakeholders will one day accept that truth and as a result learn how to perpetually re-think and re-define the way their institutions work. Continuous improvement will need to be accompanied by a willingness to replace the embedded models, interfaces, processes and yes, people, with more relevant, dynamic, and effective ones.

It is yet another scenario where our fears are our greatest enemy. The biggest challenge with adaptability, or psychologically said, amenability and capacity to change, is fear. Systems are derived from collections of people, whose actions, decisions, and behaviors determine the value-creating capacity of the system itself. For our systems to become more adaptive the practitioners within them will have to realize that the threat (and cost) of not changing is greater than the threat (and cost) of change itself. With that calculus as their motivator, they will be able to eschew fear and lower its voice in the game. (Consider this, were we to implement a Universal Basic Income program that covered survival and safety needs, more leaders of and participants in our societal systems and society might actually fear change far less.)

One example of a newly critical sub-system in America that has adopted this hyper-adaptive, fearless mentality: Amazon. With 10.4% of all 2022 U.S. retail sales and 37.8% of U.S. retail e-commerce, Amazon's phenomenal scale and omnipresence in American life can be partly attributed to what the company calls "Day 1." Day 1 is Amazon's motto, succinctly capturing its commitment to customer focus and constant adaptation, a view first declared by founder Jeff Bezos and explained by Daniel Slater, Worldwide Lead, Culture of Innovation, at Amazon Web Services:

> Day 1 is both a culture and an operating model that puts the customer at the center of everything Amazon does. We strive to deeply understand customers and work backwards from their pain points to rapidly develop innovations that create meaningful solutions in their lives. Day 1 is about being constantly curious, nimble, and experimental. It means being brave enough to fail if it means that by applying lessons learnt, we can better surprise and delight customers in the future.
>
> Contrast this to a "Day 2" mentality: as a company grows over time, it needs to adjust its approach to effectively manage the organization as it scales. The danger is that as this happens, decision-making can slow down, and the company can become less agile, moving further and further away from the customer as it rotates focus towards internal challenges rather than external customer-centric innovation.
>
> This doesn't happen overnight; it can creep in gradually, and manifest in small ways which on their own are not immediately alarming, or even readily apparent. If left unchecked, a Day 2 mentality can manifest. When asked "What does Day 2 look like?" Bezos, in his 2016 Shareholder Letter, answered: "Day 2 is stasis. Followed by irrelevance. Followed by excruciating, painful decline. Followed by death. And that is why it is always Day 1." To avoid Day 2 culture, a company must be hyper-vigilant, remained focused on its customers, and stave off practices that hamper its ability to rapidly innovate.

The Adaptation Imperative in Education

Many of the world's foundation systems are lagging because of a Day 2, entrenched, status quo culture, but none worse than education, the system that is arguably the most critical to the future of humankind. Without a Day 1 education mentality that dynamically adapts to changing world demands and actively stewards new technologies, the gap will only grow. Without a Day 1 pedagogy focused on our expanded measures of human progress, our citizens will continue to fall behind. Without a Day 1 commitment to significantly increasing our efforts to teach adaptive capacities and principled moral fortitude more and more lives will go unrealized, the divides will grow, and the collective condition will suffer. The ability to stave off the threats, to repair the divisions, and to close the gap, is almost entirely predicated on how we prepare ourselves and future generations to adapt, manage, and lead in a world that is moving far faster than we are. And if we are simply frustrated with the actions and beliefs of people around the world and next door, the best place to look for a root cause is the education system that helped develop them. Our education system is not teaching people what they need to know or how they should behave.

The higher education system in the United States is the poster child for the disconnect between what we need and what the system delivers. Its longstanding, centuries-old orientation has been to educate, and its primary measures of success have been graduation rates and the required societal stamp of approval: academic degrees. It was a system originally designed for the elite, but slowly democratized to accommodate more and more of the masses. Even though the system was opened to support a more socio-economically diverse population, its intentions and core methods were never reset or re-calibrated to accommodate these new types of "customers," their real-world needs and expectations, or the rapidly changing societal dynamics they would face upon graduation. While most liberal arts institutions continue to hold on to a "life of the mind" purpose, largely achieved (until the pandemic) via physical classroom-delivered monologues and multiple-choice testing for the price tag of $60,000 per year, the pressure is mounting on colleges and universities to respond, to change. Specifically, they are being challenged to broaden the scope of their institutions' intentions, innovate their systems, known as pedagogies and teaching methodologies, while reducing delivery costs, all directives reflected in the introductory statement of a 2017 research paper written by Alana Dunagan of the Christensen Institute, an innovation think tank:

Higher education leaders today confront a bevy of criticisms ranging from worsening affordability and persistent socioeconomic disparities to a lack of relevance in the ever-changing economy. Institutions are beset by internal challenges and external pressures. Business models are cracking under enormous pressure as state appropriations decline and net tuition growth wanes. Business as usual simply can't continue.

Alongside Dunagan's use of the apt and familiar phrase "business as usual" should be the apt and familiar metaphor of colleges and universities operating once again as ivory towers, highly siloed, insular entities with little connection to how the world works, how it should work, or what their customers want. It's a reality called out by the World Bank in a 2019 article titled "The Education Crisis: Being in School is Not the Same as Learning" and captured here:

> The world is facing a learning crisis. *While countries have significantly increased access to education, being in school isn't the same thing as learning.* Worldwide, hundreds of millions of children reach young adulthood without even the most basic skills like calculating the correct change from a transaction, reading a doctor's instructions, or understanding a bus schedule—let alone building a fulfilling career or educating their children.

The learning crisis applies to developing and developed countries and every level of education, from the impenetrable bastion of postsecondary education and its inability to adeptly respond to new market expectations, to the rigidity of most primary and secondary school programs. The pandemic made the situation and need for rapid adaptation that much more dire. A joint study by UNESCO, UNICEF, and the World Bank in July 2020 found that as many as 463 million students around the world were not able to "go to school" during the pandemic because of a "lack of remote learning policies or a lack of equipment needed for learning at home." In 2 short years the pandemic delivered a decade's worth of "momentum for change." If we're intentional and adaptive enough, we should be able to take advantage of what it has exposed and the motivation it has presented to us.

Most schools across the globe continue to manifest a face-to-face, brick and mortar "analog" versus digital bias, including a legacy focus on degrees, credits, certificates, and the passing of tests. Looked at through the lens of human progress, traditional assessment models should drop grades as the primary measure and instead gauge whole human development and the acquisition of the skills and sensibilities for today's world and work. The gap between what educators continue to think are sufficient outcomes versus what employers (and students themselves) expect was captured in a 2013 study by the Association of American Colleges and Universities. When asked the question, 93% of employers agreed that "a candidate's demonstrated capacity to think critically, communicate clearly, and solve complex problems is more important than their undergraduate major." That same study underscored the importance of real-world ready capacities over rote learning:

> Across many areas tested, employers strongly endorse educational practices that involve students in active, effortful work—practices including collaborative problem-solving, internships, research, senior projects, and community engagements. Employers consistently rank outcomes and practices that involve application of skills over acquisition of discrete bodies of knowledge. They also strongly endorse practices that require students to demonstrate both acquisition of knowledge and its application.

The shortfalls in the traditional, formal education system are not new. In 1996 the European Commission, the executive branch of the European Union, commissioned a study to take an objective look at the overall education system and make recommendations on how to advance it. The Delors Report, named after Jacques Delors, the President of the European Commission, challenged the integrity of the traditional system suggesting that its longstanding curricular emphases missed significant aspects of human development. It proposed instead a new construct that melded the concept of "learning throughout life" with four distinct but overlapping pillars of learning:

Pillar I: learning to know, giving all learners the base of knowledge that they need and the motivated curiosity to know more.

Pillar II: learning to do, imbuing the ability to translate theories and skills into the capacity to solve problems, convert opportunities and collaborate with others.

Pillar III: learning to be, framing the path of learning who one is, to embrace that finding, and to accept the autonomy, judgment and responsibilities that come with it.

Pillar IV: learning to live together, motivating the recognition and appreciation that we are all different and yet entirely inter-dependent.

To my mind, they remain four essential pillars of being human and the learning required to realize the full potential of a life and our collective existence. And yet, 25 years later, few schools have adapted by adopting them as their purpose or pedagogical underpinning. In writing about the Delors recommendations, Alexandra Draxler, once UNESCO's Director of the Task Force on Education for the Twenty-first Century, called out the disparate intentions of the then current system with what the Delors Report was calling for:

> The value of (traditional) education was widely expressed in "rates of return" on investment (public and private) and helped reinforce a strongly capitalistic and productivistic view of the value of education. It (the Delors Report) proclaimed a deeply humanistic vision of education as a holistic process, linking the acquisition of knowledge to practice, and balancing individual with collective competence. It posited the fundamental and idealistic view of education as much broader than economics.

Both inside and outside the sclerotic, almost monochromatic formal education system there are colorful if sporadic glimmers of human progress hope. Alternative models are emerging that reflect aspects of the Delors construct and the whole human development imperative. Some are long-established programs like Montessori schools, a method of childhood education that eschews discrete testing and replaces it with experiential learning environments that respect and support "free will." The model's essential tenets include mixed-age classrooms, student choice, project-based learning, and whole child consideration, with teachers serving both as teachers and facilitators of each student's growth, all likely improvements on the traditional methods and system of teaching. Today out of an estimated 3 million primary schools in the world, less than 1% subscribe to some aspect of the Montessori method. While adoption has been increasing of late, the reluctance of the traditional system to embrace any of the tenets exemplifies the reluctance of the system to change.

At the other end of the spectrum are a select group of colleges and universities that have chosen to move away from the status quo and replace it with more innovative, real-world ready approaches. In the New England area, considered by many the hub of U.S. higher education, there are several adaptive players. Northeastern University in Boston, Massachusetts has designed a 5-year undergraduate program that incorporates a full year of work-study to prepare its graduates for the workforce. Worcester Polytechnic Institute has made its focus project-based with interdisciplinary learning, while Southern New Hampshire University has been ahead of the student access curve via its innovative commitment to online learning. As some of these traditional higher education institutions attempt to re-make themselves and their relevance, alternative models are also beginning to gain traction, including Minerva University, a hybrid, global institution that eschews brick and mortar classroom buildings and focuses on experiential learning and global citizenry.

Alternatives continue to pop up on every stage of the education path and in many countries, offering preschool programs to continuous adult education models. Exemplars range from the Khan Academy and the Waldorf School in America, the Invention Convention in Latin America and LearnLife in Spain. One of the most compelling new entrants is School of Humanity, a globally accessible hybrid and online High School focused on preparing students to become future-ready by imbuing them with the requisite competencies, mindsets, and behaviors needed to live lives of fulfillment and meaning while contributing to society at large. The program is effectively an innovative amalgam of best practices, derived from the latest educational thinking and other progressive 21st century models that have proven the superior value of outcome-based, interdisciplinary and project-based learning delivered within a global context. One of the school's most distinct and popular courses is *Flourishing*. A requirement, *Flourishing* is designed to help all learners find a greater sense of meaning and self-confidence while expanding their understanding of who they are and how they can both be more present and actively participate in the world at large. Every grade and level of every school in the world should include a *Flourishing* module, including what corporations "teach."

While there is hope in the emergence of programs like the School of Humanity and the growing number of alternatives, there remains the question of whether the 800-pound gorilla, the traditional system of education, will adapt fast enough to survive or better fulfill its obligation to holistically prepare students for what lies ahead. No one knows exactly how many schools there are in the world, with estimates ranging from a few million to many million. But what is known is that billions of children and young adults go to school every day to places that are reluctant to change, and they are places that must change to teach their students how to adapt to change. It is a reluctance that is resulting in more and more Americans questioning the value of a college education as evidenced by a study conducted by Strada Education in 2021. Their survey revealed an almost 20% drop in young people believing that higher education was worth the cost.

As Paine called for universal basic income in 1795, today's revolutionaries must call for a wholesale re-invention of the most important system we have, education. Business leaders and parents will need to step forward to question how these institutions are being operated, the methods they are using and what outcomes they are seeking. And they must demand change. The leaders of the institutions themselves will need to step forward to break the reliance on change-averse, recalcitrant faculty, and teaching unions or at least negotiate their willingness to rapidly evolve the value proposition, from pedagogy and curriculum to delivery options and costs. We, as leaders in our communities and families will come to reject the alarming difference between what the teachers of our children make ($66,000 on average in America in 2022) and a country's sports star's average package ($2.6 to $8.3 million in America in 2020) and demand more education funding from more sources now. Of note, in 2023 the United States spent approximately 6% of its annual GDP on education, ranking it number 66 out of 198 countries tabulated. For the world's reigning superpower, that metric may be a portent of a less superpower future. It's also a consistent marker of an empire entering its final stage: Glubb's Age of Decadence.

Employers as Educators

As the formal education system becomes more intentional, dynamically adaptive, and responsive to externalities and "customers" changing requirements, so too will employers as educators. The long-running pandemic exposed many truths, ranging from the ability of white-collar workers to be productive from home to how they feel about the job they are being asked to do. It all suggests that as more and more companies and organizations promulgate a purpose beyond profit and express a commitment to more active contribution to society, they must also embrace their responsibility to help their employees evolve to better pre-pare for both what is and what will likely be next. Corporate "training" budgets traditionally managed by Human Resources and focused on purely functional and management skills need to be replaced by a unilat-eral commitment to "whole employee development," including programs focused on strengthening employees' well-being, their mental, spiritual, financial, and relationship health while imbuing them with the capacity to constantly adapt and re-skill themselves. The corporate commitment to re-skilling is growing, as witness Amazon's 2019 announcement of its plan to re-train a third of its U.S. workforce, some 100,000 workers and the invest-ment of over $700 million, or roughly $7,000 per employee. To their credit, Amazon's re-skilling commitment goes beyond their own self-interest:

> While many of our employees want to build their careers here, for others it might be a steppingstone to different aspirations," said Beth Galetti, a senior vice president of human resources at Amazon. "We think it's important to invest in our employees, and to help them gain new skills and create more professional options for themselves. With this pledge, we're committing to support 100,000 Amazonians in getting the skills to make the next step in their careers.

The caveat to Amazon's human development largesse is this: re-skilling programs across the planet need to focus not only on functional skills but on teaching employees how to re-skill themselves, how to embrace change, and, of their own volition, how to adapt. We need to teach them and each other how to fish. Because without the motivation and ability of the indi-vidual to actively re-skill themselves, the aggregate re-skilling task and costs now carried solely by employers will eventually become unsustain-able, prompting a more concerted turn towards automation and robotics as the more cost-effective labor solution. Whether an employee plans to stay with the employer for 3 years or 30, the employer and the employee must work on their adaptive skills and future readiness in every human form.

In her *Forbes* article "Reinvention Mandate: Succeeding in 2020 and Beyond" Julie Jungalwala, the Founder and Director of the Institute for Future Learning further explains,

> While these (reskilling) efforts are critically important, they are not sufficient. Training and skills inventories alone will not sustain

the level and frequency of skilling required. Instead, we need to put people's ability to grow, change and evolve at the center of the skilling process. Meeting future workforce needs requires an in-depth understanding of the human side of reskilling and upskilling: We must decode the black box of human reinvention.

She goes on to cite the work of Harvard Professor Robert Kegan and his theorized three stages of human development: the socialized mind, the self-authoring mind, and finally the self-transformational mind, a model that echoes Maslow's three highest levels of need: belonging, self-esteem, and self-actualization. These stages of growth are neither better nor worse, but as Jungalwala so aptly sums up,

> Humans have evolved as long as we have walked the planet. What is different this time around is that reinventing ourselves is no longer optional—living and working in (today's world) and beyond mandates reinvention.

THE THIRD I: INTEGRATION

Imagine a day where our core systems have clearer, more relevant, and human intentions, intentions that directly contribute to our collective well-being and latitude. And imagine that these systems have learned how to continuously adapt to deliver on those outcomes. The world, our world, would be all set, right? Not quite. Because without the third "I," without the seamless integration of technologies, data, processes, and interfaces our systems and institutions will stumble in their efforts to nimbly respond to changing market requirements, competitive threats, and disruptive alternatives. They simply will not be able to realize their newfound intentions without integration, without far greater internal and external connectivity. In an infinite world increasingly fueled by less and less friction, where all components and action are implicitly and explicitly linked to a multitude of other components and actions, we should all let go of the notion that one thing can be everything. The education system alone cannot prepare our children and employees for what is next. The health care system alone is not responsible for our health. The financial system alone is not respon-sible for our financial future. One thing, one system, cannot be everything. Everything requires a collaboration of collective systems and individual contributions. This is the definition of what is called "systems thinking," a holistic approach that dictates that integration within a system should be directly considerate of other external systems. As they must be linked, they must also be aligned. Jim Collins, the author of *Good to Great* and *Built to Last* once wrote,

> Building a visionary company requires one percent vision and 99 percent alignment.

His declaration equally applies to the task of building visionary core systems. Alignment is essential. Consider the two highest performing "systems" in the world: planet Earth and the human anatomy. Both are remarkably aligned, inter-connected systems where a mind-blowing array of components are all linked. Something that happens on the North Pole impacts things happening on the South Pole. It is a mutually reliant eco-system of ecosystems (the climate crisis is a stark reminder of this). Our bodies are the same. When we stub our toe, our brain knows it. And when our brain is depressed, our body shows it. There are no discrete elements or actions in systems like these. They are perfectly configured and linked, designed to produce healthy outcomes, distinct capacities, and sustainable value.

Not coincidentally, humans have created a means to achieve deep integration within the software development world. It's called an Application Program Interface, or API. An API enables two distinct technology systems to interact and communicate with each other without negatively impacting the core functionality of either system. Data, inputs, and outputs can be delivered in a standard way without causing errors or diminishing performance. The concept of defining and building stable, additive interfaces to elevate the value of the combined system is not new. Applying that concept to our core societal systems, both within and across them is. And it begins again with clear intentions, i.e., what does each system participant want and need from the other and how will the systems whole yield greater value than the sum of its parts. The answers to those seemingly simple questions are complex, but also key to our revolutionary march towards greater human progress.

Integration: A Health Care Imperative

The health care system is perhaps the best example of a core system that struggles mightily with both inter- and intra-connectivity, resulting in soaring costs, reduced quality of care and patient safety, and for some, a lesser quality of life. Unlike the traditional education system, the majority of the traditional global health care system players acknowledge the need to adapt and specifically to deliver a far more integrated system of care. The research being conducted on the topic is extensive, the experiments are many, and yet the progress is painfully slow. In America today, most large hospital systems are struggling with the foundation principles of effective integration. The technologies in the operating and examination rooms don't talk to each other. The patient profiling systems are many and redundant, and in most cases still paper-based. The ability to easily combine traditional offline examination and treatment with online assessment and follow up is riddled with holes, hurdles, and dead ends. (A sliver of progress: as with education, the pandemic forced the system to finally take online delivery (tele-health) seriously.) Data in health care systems is not shared seamlessly by department and the entire standard of care is focused on treating the discrete condition versus understanding the whole picture and then treating the whole human. And perhaps most egregiously, the ability to easily connect the dots between the care provided and the associated (reasonable) costs incurred is virtually non-existent because of a legacy payment system that is opaque, archaic, and arguably designed to cover the sins and exorbitant costs of a completely disintegrated system. Consider the global health care system's failed response to the pandemic. From being caught flat footed on the basic issues of inadequate PPE supplies and vaccine access to the overwhelming burnout of care givers, the impact on the system because of a lack of systems integration was devastating.

The root cause of the lack of integration and connectivity in health care lies in a familiar combination of a cacophonous and constant flood of new technologies, grandfathered complexities, and unhelpful human behaviors, all legacy issues plaguing the other core systems and humankind in general. The technology is often ahead of our ability to completely understand it or to know how to best apply it. Our all too human tendency is to look at problems and the potential solutions as discrete topics when everything is linked. We succumb to our behavioral and budgeting biases favoring remediation, waiting for the problem to demand a low-cost solution versus investing in prevention. We reject alternative methods as potentially effective because of our lack of familiarity or resistance to change. And finally, we are unwilling to accept the possibility that the mind, body, and spirit connection is real and relevant to the treatment of a disease or chronic condition. The list of behavioral and capacity-based integration roadblocks goes on and on.

In their research paper "The Need for Systems Integration in Health Care" presented in the *Journal of the American Medical Association* in 2011, Simon Matthews and Dr. Peter Provonost underscored the system integration problem with some telling questions:

> Why are clinicians, who do not necessarily have a background in technology or human factors engineering, painstakingly deciding and purchasing which pieces of equipment should be cobbled together in an emergency department, an ICU, or an OR? Why are clinicians from individual hospitals trying to negotiate the lowest prices for individual pieces of equipment? Why are clinicians also responsible for ensuring that these systems communicate and work concurrently when there was no forethought by equipment manufacturers to help meet these goals? Clinicians and administrators are trying to build hospitals piecemeal, buying technologies one by one, hoping to make equipment and technology talk to each other. Yet in so doing, they are increasing health care costs and reducing health care quality.

Thirteen years later we are not much further along. The authors went on to theorize that the answer to the systems integration question in health care lies in creating a systems integration capacity outside of the system itself. They proposed the health care equivalent of a Boeing, the largest manufacturer of planes in the world but a system integrator at its core. Their compelling point was and is that the health care system is so complex, so riddled with disconnects and operated by biased professionals who lack systems integration knowhow, that the only way to achieve seamless integration is to outsource the function to objective experts who can design and build the new system from scratch. Extending the Boeing analogy, Mr. Matthews, and Dr. Provonost are suggesting that one cannot transform a plane with propellers into a fighter jet while in mid-air. Recognizing the unlikelihood of the health care industry accepting an external for-profit systems integrator as the answer, they propose an alternative path: bringing together university-based academic medical centers and their public health and school of engineering counterparts as integration learning co-laboratories to design and test integrative models that bring together the right mix of protocols, process, technologies, and behaviors to deliver quality care at a lower cost. At first glance it seems like a viable idea, less so if we consider the non-adaptive, innovation reluctance of the higher education system itself. With their collaboration path proffered as the answer, the authors conclude with a cautionary note, a statement that implicitly calls for more Humanist Revolutionary leaders to step forward:

> Although the technology is available to create an integrated OR, ICU, or hospital, the leadership to create the needed partnerships has not emerged. It will take leadership in academic medicine and private industry to convert early successes from partnerships into mass market solutions that meet clinicians' needs, improve safety, and reduce costs of care.

Implicit in any future solution is the need for API capacity, the ability of entities both inside and outside the system to easily "plug in" to add value or derive value from each other. That "open source" notion points to the fourth "I" in the framework, inclusion.

THE FOURTH I: INCLUSION AS ACCESSIBILITY

As platitudinal as this is, it's inarguable that all societal systems fail at contributing to collective human progress if they are not available for all humans to participate in and benefit from. Inclusion and accessibility are essential in a humanist society that cares about a shared future. And yet inaccessibility exists today in many forms. Humans are barred from access to education because of economic and geographic circumstance, from job opportunities because of race, sex, or educational background, from health care because of physical access or financial means, from career advantage because of limited networks, and from program access because of physical and digital abilities. The list of unfair access issues is unacceptably long. And without access to the basics of education, health care, and financial means, lives cannot be fully realized, and greater human well-being and latitude is a pipe dream. For all our advances as a civilization, major accessibility issues within our core systems remain, with the global financial system being at the top of the list.

Inclusion: A Financial System Imperative

As of today, as many as 2 billion people, roughly 25% of the world's population are "unbanked" (6% of the U.S. population), which means they cannot access basic financial services, including checking, savings, and lending, because they lack the required identification and documented credit history. Thus, without financial inclusion they are forced to use expensive and unreliable alternative systems like paycheck-based loans and third-party check cashers. According to David Rothstein at the Cities for Financial Empowerment Fund, the cost of those alternatives can add up to as much as $40,000 over the lifetime of an unbanked individual. But worse, without access to basic services, the unbanked remain economically unequal, far less able to attend college, forge entrepreneurial paths, or simply invest in their own futures.

Efforts to reduce the number of unbanked are coming from the private, public, and NGO sectors, including the World Bank's Universal Financial Access 2020 initiative that began in 2015, targeting helping one billion people gain basic banking access. One of the more innovative programs has been a for-profit enterprise called M-Pesa. Launched in Kenya as a joint venture between Vodafone and Safaricom in 2007, M-Pesa is a cell phone-based money movement service, a bank without being a bank. Since its inception over 17 million individuals in multiple countries in Africa have become "banked" and with that gained access to greater opportunity.

Thanks to innovations like M-Pesa, financial system access is improving. But that progress comes with a qualifier and a reminder that core systems need to be in synch with one another for the whole to be greater than the sum of the parts and for the outcome to be bottom line beneficial to all. As the Humanist Revolution seeks to motivate financial system access to the remaining 2 billion plus unbanked humans, we should also make sure that the push for financial inclusion is accompanied by systemic investments in financial education. As we seek to equip more people with the ability to obtain credit and to invest, we must make sure they understand the accompanying risks. Without financial education and literacy, expanding financial inclusion can quickly become the sub-prime mortgage debacle of 2008, one of the main drivers of the global economic crash and the cause of misery for millions of people.

(The idea of Universal Basic Income can and should also be considered part of the financial system access solution. By providing all humans with the ability to meet their basic Maslovian needs, we free them to realize their potential and to become contributing members of the world. It's a case where providing the system ends can be as powerful as providing the system means.)

THE FIFTH I: INFINITE TIME HORIZONS

The fifth "I" criterion is the need to change the time horizon of the core systems, including democratic political systems. All our systems need to be re-built to perpetually adapt. And their intentionality must be directly considerate of their customers' time horizons. We are far from this ideal. The traditional education system is designed to end at high school (secondary) for most, college, or university for some. The current time horizon for formal, structured, and generally required learning is 12 to 16 years. Given our new perpetual adaptation mandate, it must be for life. Because the health care system's intention is to cure and not to prevent, it operates on a time horizon of days, weeks, and sometime months, whatever time it takes to get the patient out of the system. The time horizon of the health care system should also be for life. The time horizon of the financial system has followed a similar path, enabling consumers and businesses in the here and now, with little consideration of the downstream consequences and how best to help them prepare for later in life challenges and demands. The stock market time horizon is not a life or even a day but instead a millisecond, just enough to enable automated high-frequency trading. In the public markets, quarterly earnings reports are considered long-term and taking a "long position" means holding on to the asset until after lunch.

But I propose that in much of the Western world it is the democratic political system that most calls for a significant change in time horizon. It is the political system, and typically the legislative and executive branches of government that produce most of the policies and directives that determine the trajectory of a democratic nation. In the United States the average tenure of an elected president over the last 100 years has been approximately 5 and a half years. The average tenure of a congressman has been 8 years, and that of a U.S. Senator 11 years. Juxtapose those time horizons with the time horizons required to take on the growing number of existential threats, beginning with climate change. Repairing the damage technology and human behavior have done over decades will require bold, sustained policies and decisions that stick for decades. If the leaders are replaced every 5 to 11 years, and the new leaders opt to dismantle the actions of their predecessors, how can long-term progress be made in general and against the threats in particular? The leadership time horizon can be elongated not by extending term limits but by changing the paradigm of rule and shifting the transformative leadership responsibilities to the individuals and entities that can participate over time. Our clarity and consensus agreement regarding our measures of human progress would then naturally interact with the political system, assuming our governing institutions carry responsibility for our collective interests. It's an assumption that Robert Bellah and his co-authors of *The Good Society* present as an imperative:

Indeed, the great classic criteria of a good society—peace, prosperity, freedom, justice—all depend today on a new experiment in democracy, a newly extended and enhanced set of democratic institutions, within which we citizens can better discern what we really want and what we ought to want to sustain a good life on this planet for ourselves and generations to come.

This synergy between the voice of the people and the rule of our democratic institutions is a critical consequence of the revolution, resulting in sustainable policies that will no doubt take decades to form.

THE SIXTH I: INCENTIVES

The sixth criterion required to ensure the ongoing viability of our overhauled core systems is the matter of building in the right incentives. Incentives are central to how *all* systems work. Incentives motivate or de-motivate individual and collective system engagement, whether the humans are suppliers in a system or its customers. Incentives encourage us to conform to what a system requires, which is on occasion collectively useful, and as often, not. Most fundamentally, incentives are the ways that systems scale, by harnessing self-interest for collective purposes. And across the board, the incentives in our core systems appear out of alignment with the new requirements and challenges of the world. In higher education, the faculty's incentive of tenure is often attached not to graduate readiness or some measure of improved human capacities, but to how often a professor has published. In health care, financial incentives are usually attached solely to remediation and short-term patient satisfaction with little reward for preventative efforts or impact. The primary political system incentives appear to be some unhealthy combination of power gained and partisan followers appeased, incentives that are getting in the way of our leaders' willingness to compromise and care for the whole country. The financial system has suffered from incentives rewarding volume at any cost, short-term performance, and motivating mutual back scratching among the inner circle of banks, financial firms, and credit rating agencies. At a 2018 U.S. Chamber of Commerce meeting William Dudley, then the President and CEO of the Federal Reserve Bank of New York, gave a talk entitled "The Importance of Incentives in Ensuring a Resilient and Robust Financial System." In referring to financial system abuses that happened long after the 2008 crisis clean up, he stated,

These recent cases were particularly disturbing in terms of their scale and flagrancy, and—in the case of the rate-rigging scandals— the collusion by employees across firms. I am particularly struck by the fact that manipulation of the foreign exchange market occurred even after the LIBOR scandal was well known. These episodes underscore the tremendous power that incentives have to influence and distort behavior, potentially leading to massive damage to bank cultures, reputations, and finances.

Incentives drive behaviors, and the core system that may be most impacted by the wrong or at least not ideal incentives is that of a capitalist economy. Capitalism is about the incentive of individual reward and the premise that the harder one works, the more ingenious one is, the more likely one will benefit both extrinsically (capital gain) and intrinsically (social capital gain). Remember Will Storr's *The Status Game*. In theory, the drive of the individual coupled with the demand–supply curve allow the economy to optimize, to self-regulate, and require far less government intervention. The problem is that the dominant incentive of individual reward overwhelms the concern for collective good, often fueling economic inequality and resulting in a society of haves and have nots. Capitalist economies are also marked by cycles of boom and bust, as consumer sentiment and capital market sensitivities can trigger massive, sudden shifts in buying behavior. On paper, and to some degree in practice, socialist economies are designed to solve all that. Unlike capitalism, which is focused on Maslow's need states of self-esteem and a perverse form of self-actualization, socialism is oriented towards Maslow's more base levels of sustenance, safety, security, and belonging. The incentive is not individual reward but rather a collective safety net. Even with the good of the whole as their primary focus, socialist economies often struggle to deliver on the promise however, lacking the motivation to innovate, and suffering from the bouts of corruption and system abuses that get in the way of healthy growth, fair distribution, and the ability to respond efficiently to external market conditions.

While the economic systems' choice appears binary, it is clearly not. What is obvious is that the incentive structure of an increasingly globalized economy powered by capitalism should be re-considered, with a balance found between the motivation of individual gain and the need for collective benefit. According to the 2022 Credit Suisse Report, in 2020 1.2% of the total adult population held 47.8% of total global wealth, approximately $221.6 trillion. 53.2% of the adult population carried just 1% of the total. Without refined and more considerate incentives this extreme division of global wealth will only grow, until such a time that the have nots have had enough. And then a not so Humanist Revolution will likely begin.

An Invisible System

As we act on re-focusing, re-building, and potentially replacing our core systems we must also address an invisible but structurally central system, the system of technologies. At times an enabler, an essential integrator, and occasional trouble maker, it is a system that has no consensus intentions, few if any formal rules or shared protocols, and no designated guides or stewards. Outside of education it may be the most important system upon which to re-focus and re-build.

The system of technologies and its outputs have found their way into the center of every other system. And yet it appears a system out of control. Whether it's education's awkward adoption of online technologies during COVID, or health care's jumbled array of technologies that don't talk to each other, or the finance system's head scratching regarding what to do about decentralized finance and cryptocurrencies, the technology system is driving speed and scale in all of them. It's operating like a bull in a china shop, busting things up and not clear on what exactly it's after. And the current riders of that bull are mostly the technology companies themselves, nominally being held accountable for the damage the bull is doing. Add to the broken china the global growing issues regarding data privacy, freedom of speech abuses, cyber hacking, misinformation proliferation, AI ethics, deep fake technologies, and fundamental Internet access inequality. We have a troubling and growing portfolio of problems calling out for a far more structured, centralized global approach to the invisible system called technologies. We need a different rider on that bull, and we need to put the bull in a controlled corral. Without such an approach, we will have allowed technology to become yet another existential threat. AI is already here.

One admirable attempt to control a portion of the system of technologies was made by the European Union in 2018 when it launched the General Data Protection Regulation (GDPR) to establish rules around governing the use of personal data. Three years later, adoption and adherence has been inconsistent with nagging concerns regarding conformance to the regulations undermining innovation and economic growth. In a 2018 speech at a privacy conference in Brussels, Apple CEO Tim Cook called out the resisters and called for the universal adoption of the GDPR:

> It is time for the rest of the world—including my home country—to follow your lead. We at Apple are in full support of a comprehensive federal privacy law in the United States. There and everywhere, it should be rooted in four essential rights: First, the right to have personal data minimized…Second, the right to knowledge… Third, the right to access…And fourth, the right to security. Security is foundational to trust and all other privacy rights. Now, there are those who would prefer I hadn't said all of that. Some oppose any form of privacy legislation. Others will endorse reform in public, and then resist and undermine it behind closed doors. They may say to you, "Our companies will never achieve technology's true potential if they are constrained with privacy regulation." But this notion isn't just wrong, it is destructive.

The troubling issue of the lack of an agreed-upon global structure and our inability to enforce a collective policy for mutual benefit are correlated. Absent a centralized, empowered structure to universally manage and guide the system of technologies, it would seem Cook's predicted path of destruction will come to fruition, a risk that many of his "techno-utopian" peers apparently deem worth taking. Another option is to somehow repeal the laws of capitalist markets and forbid self-interest, the two forces that drive companies to abuse resources (technology outputs) that they don't pay for. Given the unlikeliness of that path, the other choice is for all of us to take responsibility for understanding and managing the risks and choices as individuals. And that presents the seventh criteria in the "seven I's" systems framework:

THE SEVENTH I: INDIVIDUAL RESPONSIBILITY

As noted, it is unlikely that the technology system can re-boot itself, the education system can re-train itself, the health care system can heal itself, the political system can govern its non-partisan transformation, the finance system can decentralize itself, and the capitalist system can suddenly put the good of the whole before its survival of the fittest mantra. The promulgation for change will not come from within, and it won't come from centralized governments. As comforting as it is to think of governments as a viable solution, history suggests they are not. Over the centuries, centralized authorities have proven to be useful for setting standards and establishing compliance measures, just as the criminal justice system is the "court of last resort" in encouraging good behavior. But as discussed above, the incentives and intentions of our governing systems in the developed world are themselves out of synch with the issues and realities we face. The only effective policing mechanism, the only means to steer and control all the systems, is self-policing, by companies, communities, employees and yes, you and me. As we push for system overhaul, we will also actively support the designing, experimenting, building, testing, and scaling of completely alternative systems. We too must learn to let go of our own legacy notions that are attached to these legacy systems, and step forward as activists who will accept nothing less than substantive change not just in means but in ends.

The question of ends points to a final systemic point, and harkens back to Chapter 7, "Beginning with the End in Mind". That which is not measured cannot be improved upon. A collective set of global measures of human progress, combined with the refined intentions for each of re-designed core support systems should become our planet's performance dashboard. *The United Nation's 17 Sustainability Goals* remain relevant, but what is more actionable is gauging whether the core systems that really determine our collective well-being and our capacity to beat back our existential foes, are doing their job. Are they delivering on their human progress intentions and are they consistently manifesting the other six essential criteria? Are they adapting fast enough? Are they collaborating? Are they working towards universal inclusion? Do they have the right incentives in place? The right time horizons? Are they being pushed to change by responsible revolutionaries? And most importantly, is the system of technologies at the center of it all being proactively managed by us as an enabler versus disabler of those systems and of human well-being as a whole? Without systems change we cannot expect the outcomes to change. Which is why these are the questions to be answered and the work we must do.

Systems produce consequences.

CHAPTER 10

Changing the
Ways of Us

"Our problems in technology, in politics, wherever, are human problems. From the Garden of Eden to today, it's our humanity that got us into this mess and it's our humanity that's going to have to get us out."

Tim Cook, Apple CEO

We are slowly realizing that the ways things are is increasingly disconnected from our latent hopes for ourselves and our species. But as uncomfortable a picture as that is, there is another equally discomforting truth: the way things are is a direct consequence of the way we are. We cannot achieve greater collective well-being without first looking in the mirror and choosing a different image of ourselves and our associated behaviors. The systems work we need to do, the ability to change the paradigm of rule, to achieve our whole human definition of progress, require deep changes in the way humans think and behave. Policies alone will not do it. New governing structures will not do it. Even newly constituted systems that follow the seven I's won't get us there either.

For those changes to happen, we need to change. Our revolution is a push for the wholesale transformation of our ways of being, managing and leading by putting our human truths and priorities before all else, beginning with technology. And it is not a new proposition. The humanist movement began centuries ago. There is a long history of individuals and organizations attempting to reset the definition of human intent and human ways of being. A history of intelligent, caring people pushing back against the

 DOI: 10.1201/9781003089902-14

legacy construct of religions, of oppressive regimes, and the omnipresent enemy known as self-interest. From the ancient philosophers, including Petrarch and Erasmus, to more modern manifestations like Bertrand Russell, they were all in their way attempting to elevate the understanding of us, to redefine the intent of us, and to motivate others to behave differently towards us. All of us, and all living things. They knew, and we know, that the solution to our problems is both mechanistic and humanistic. And without enough of the latter, the mechanics will fail. The techno-utopians are wrong. Without far more humanistic considerations at the fore, technology will drive us and our planet off a cliff. The difference between the humanistic work of Petrarch and ours is this: the stakes are now existential.

THE ROOT CAUSE

Mindsets and behaviors are the root of all actions, all decision-making and all consequences, and therefore the biggest determinant of our future. As written by Niccolò Machiavelli, the 16th century philosopher and considered father of modern political science:

> Whosoever desires constant success must change his conduct with the times.

More recently expressed by the polymath and iconoclastic innovator Steven Titus Smith in his book, *I Am Gravity:*

> Behavior is the building block of culture, and mentality is the architect of behavior.

The way any entity creates, produces, and destroys is most impacted by its conduct, its behaviors. In this case, "its" means "ours" and that truth applies to individuals, families, cities, companies, and countries. The many consequences of hyper-consumerism and addictions to excess, the growing divides, the erosion of trust, and the rise of misinformation are at their core fueled by technology and our ensuing problematic all-too-human behaviors. The summary headline is this:

> We have allowed technology and other unguided forces to foster our bad behaviors. It is time for our good behaviors to foster technology and the world.

The author and Buddhist practitioner Jack Kornfield once wrote,

> When we consider creating the best future for humanity, the principles for a wise society and a wise life are simple and universal: Actions based on greed, hatred, disrespect, and ignorance inevitably lead to suffering. And actions based on their opposites—generosity, love, respect, and wisdom—lead to happiness and well-being. That is true for us humans, and it applies to all the technologies we develop and employ.

All outcomes are the manifestation of our mindsets and behaviors.

The decline of the British Empire and global videoconferencing giant Skype's stunning market share drop from 32.4% in 2020 to 1.35% in 2023, are both stories of consequences fueled by the wrong motivations and behaviors, behaviors that led to significant loss. And they are both examples of "The Tragedy of the Commons," an economic observation first espoused in 1833 by William Forster Lloyd, a political economist at Oxford University. Lloyd's theory, which continues to be proven daily across the world, is that the sharing of common resources inevitably results in resource depletion, as individuals (including companies and countries) are motivated to take as much as they can in the short-term without consideration of the long-term impact on the whole. Commons tragedies range from the commercial to the environmental, from the global to the local. Climate change is the tragedy of the commons writ large. So too is fast fashion, over-fishing, pollution, freshwater depletion, the addiction to social media, and most of the other problems we humans face. They are all the consequence of self-interest and the dark behaviors it motivates. The only way to avoid the tragedies of the commons is for more humans to decide not to succumb to self-interest and instead adopt a different set of behaviors. Behaviors that put the priorities of the commons now and in the future before their own and by acknowledging and valuing the self-interest gains inherent in common progress. It is a concept perfectly expressed by the 19th century author and social philosopher Ralph Waldo Emerson, as "the we of me."

TAKING THE WHEEL

Taking the wheel means more of us stepping into leadership roles and embracing new ways of behaving, guiding, and managing both organizations and the world at large to drive essential change. Behavioral adaptation by us is now our most critical competency and the only path to a sustainable future. Like young children, employees and citizens often mirror the behavior of the adults that lead them. Bad behaviors beget bad behaviors. We need more explicit and implicit leaders to exhibit good behaviors, behaviors focused on the right intentions, aided by re-designed systems, and motivated by the right beliefs and incentives—beliefs and incentives that differ dramatically from what most humans are exhibiting and responding to today. Whether more of us do little or big things differently, *anything* that is oriented towards positive transformation and the collective good is a good thing. Our actions and behaviors can and will make the difference as captured in a *New York Times* interview with Katharine Hayhoe, chief scientist for the Nature Conservancy when she stated,

> …our personal actions do matter, but they matter because they can change others. When we take that extra step of saying, "Hey, I tried a Beyond Meat burger, and it was delicious. Let's go to this place for lunch and give it a try together." If you take those kinds of actions, all of a sudden you've got 30, 50, 100 more people whose hands are on the boulder beside you, and you realize, hey, we might have a shot at fixing this.

OUR LEADERSHIP AND THE SEVEN I'S

It is no coincidence that the success criteria for transformative humanist leadership is identical to the seven "I's" criteria to be applied to systems (intentionality, innovation, integration, inclusion, infinite time horizons, aligned incentives, and individual responsibility). Given that leaders at every level of society are integrated components in the world's overall "operating system," it only makes sense that they mirror and manifest these same attributes. As with the systems, the seven criteria are linked, which is to say that for revolutionary leaders to exhibit one, we must exhibit them all. And we can start by completely re-setting the intentions for ourselves and the organizations, communities, and families we lead, including our shared acceptance of collective well-being as our definition of human progress. Imagine an agreement across the world that human well-being, deeper human connection, and the latitude of choice are the metrics of the human experiment that matter the most. Once embraced, that intention and its subordinate system measures would immediately require the leaders of all corporations and organizations to aim far beyond their balance sheets and mission statements. The current trending declarations of corporations becoming "purpose-driven" and "stakeholder-driven" would take on more honest, holistic, and humanistic definitions, and be accompanied by a re-doubled effort to meet their ESG (Environmental, Social, and Governance) commitments. In a March 2021 *Harvard Business Review* article "An ESG Reckoning is Coming," authors Michael O'Leary and Warren Valdmanis call out the gulf between what the Global 2000 have claimed they will do vis-a-vis ESG and their social capitalism intentions vis-a-vis what they have done. That gulf of under delivery or worse disingenuity is fueling a growing skepticism and discontent among both investors and customers. O'Leary and Valdmanis write,

> As evidence of the connection between a company's social and environmental impact and its financial performance continues to grow, companies ignore these trends at their peril. Ultimately, capitalism is nothing more than the sum of individual choices. Our economy is no more moved by an invisible hand than a Ouija board is moved by invisible spirits.
>
> But if capitalists are unable to reform capitalism, it will be reformed for them. The American public is already distrustful of big business, and only half of American adults under 40 view capitalism favorably—down from two-thirds in 2010. Companies that don't adapt will find themselves at odds with their customers, employees, investors, and regulators. Former Attorney General Loretta Lynch predicted that we will soon see legal action "that goes beyond regulatory scrutiny to environmental justice, to racial inequality and ESG more broadly."

(Of note, the regulatory system in most developed countries is another core system that requires wholesale transformation and the application of the seven I's. At some point the regulators will need to step forward to police whether corporations do or do not meet their societal, human progress commitments. In the interim, it is up to customers, employees, and investors to hold them accountable.)

Corporate actions exhibiting an updated and real commitment to an expanded purpose will mitigate the likelihood of legal action and result in fewer employees feeling like indentured servants and more like stakeholders aligned with their organization's cause. The 2021–2022 labor phenomena in America dubbed "The Great Resignation," whereby 12 million Americans quit their jobs during the post-pandemic re-opening revealed the level of dissatisfaction people have with their livelihoods and the organizations that employ them. Employees want to be part of something bigger, they want to work for employers with bigger, more human intentions that they can rally behind and contribute to. Daniel Pink, the best-selling author of numerous books on human behavior once proposed that most people seek three things in their jobs: a sense of mastery, autonomy, and purpose. Yes, paychecks still matter, but purpose matters just as much, and for many, maybe even more.

In organizations that have ignored Pink's insight, the contract has become broken, and can only be repaired with more humanistic, shared intentions that make their way onto corporate performance dashboards, reflecting not just organizational growth and fiscal health but global impact and collective well-being. The accelerating B-Corporation movement is an indicator that more and more companies around the world get this, with the number of "certified B Corps" topping 7,351, representing 161 industries and 92 countries (as of 2023). B Corps are, as explained by *Fast Company*'s Talib Visram,

> …companies that have been officially certified by nonprofit B Lab, for their commitment to not only pledging, but concretely showing, environmentally and socially beneficial business practices, public transparency, and the "legal accountability to balance profit and purpose." B-Corps-to-be must pass a 200-question assessment that judges performance across five impact areas: governance, the environment, workers, customers, and community.

In his book *Better Business: How the B Corp Movement Is Remaking Capitalism* Cornell University professor Christopher Marquis explains the trend and the need that underlies it:

> I do think that people are more aware than ever that we need a change, and the shift to a more stakeholder-driven economy can and will happen.
>
> The crises of the last year have only further magnified the huge structural problems in our economy, and they require a holistic solution. While it is regrettable that we have had to come to a crisis

point in terms of income inequality, racial justice, climate change, and the COVID-19 pandemic, I am cautiously optimistic that in the next 50 years, the pendulum will swing back to a stakeholder-driven system.

Marquis' optimism presumes that we will be willing to collaborate across industries and nations to re-focus and re-structure the economic system, beginning with a definition of what stakeholder capitalism means, how to thwart the self-interested, deal with global competition and incentivize investments that yield collective benefits. Whether our organizations become B Corps or not, as Humanist Revolutionary leaders we should re-define why business is in business through the lens of collective well-being, for humankind and for the people that we work for and who work for us.

THE POWER IN THE PEOPLE

At a community level our new measures of collective well-being and the latitude of choice, should influence material change. More and more citizens will be motivated to reconsider their actions, decision-making, and consumption as they acknowledge that their mark on the world and those around them is often not a positive, forward contributing one. It is the revolution's task to establish the intentions and systems that cause matters such as climate and sustainability to go from being a polite set of questions to ponder at a cocktail party to a mandatory set of daily actions to take. The intention of human progress at an individual level will require all of us to lead and act not from primal need or our calculating, self-serving brains but from our loving hearts, tapping into our compassion and acceptance that we are all one species, one people, intrinsically and extrinsically linked and mutually reliant on this Earth.

While the idea of an intention to operate from the heart will no doubt be met with raised eyes, it is one of the few intentions that in and of itself can change our future. Again, in the clear words of the author and Buddhist practitioner Jack Kornfield,

Whether we admit it or not, we are vulnerable beings, and the work of an open heart is demanding. But our crisis of heart requires it. To curtail violence and hate and to foster human well-being, we need to spread widely, in person and online, the trainings and tools of compassion, forgiveness, trauma healing, and nonviolent communication. By growing empathy and inner courage we expand what neuroscientists call our window of tolerance. Without this wisdom, we blame society's ills on others, whether the immigrants, the Muslims, the Communists — it's always someone else. Back in 1955, the author James Baldwin wrote, 'I imagine one of the reasons that people cling to their hate and prejudice so stubbornly is because they sense that once hate is gone, they will be forced to deal with their own pain.'

LOVE VERSUS FEAR

Many humans will quickly eschew the idea of love becoming our primary modality and instead hold on to hate and prejudice, all because they are afraid. As shared in Chapter 3, racism and the disdain for others not "like us" in part comes from a centuries-old contagion of fear related to the base level of Maslow's Hierarchy of Needs. The psychologist and political consultant Dr. Reneé Carr explains,

> When one race of persons unconsciously feels fear in response to a different race group—fears that their own level of security, importance, or control is being threatened—they will develop these defensive thoughts and behaviors. They will create exaggerated and negative beliefs about the other race to justify their actions in [an] attempt to secure their own safety and survival.

We need to feel superior to others because we fear the feeling of self-worthlessness and the risk of our survival, and perhaps, most primally, because we are, once again, wired for war. The first recorded examples of slavery occurred in 6800 BC in Mesopotamia, when the beginnings of land ownership resulted in warring tribes, with the vanquished ending up in chains as forced labor. That unfolding has continued for thousands of years across every continent as the enemy that was feared became the enemy that was enslaved. We fear those not like us, the enemy, because we believe they might take what we believe is rightfully ours. The dominant majority fears the others because we have what they rightfully want, the ability to survive and thrive. They fear us because we have more power, advantage, and control, and because history confirms that they should fear us. They fear that this inequality and injustice will never really end, and history again shows that it won't, unless most of us change our intentions and our behaviors. The only antidote to the contagion of fear, to stop the generation-to-generation transfer of racism, bigotry, and the derivative inequalities is to teach the next generation of the majority to love themselves. It is logical. When people love who they are, they are less likely to fear loss, to avoid acknowledging their pain, to be unwilling to share. When people love who they are, they no longer need to hurt others to feel better about themselves. When people love who they are, they are willing to give, to understand, and to find common ground with people who are different. They are willing to take less, to care for and share the commons.

BECOMING REVOLUTIONARY ROLE MODELS

Because most of us lack the intention of leading from the heart, modern society does not actively teach love of self, or love of others. In his book *Love and Compassion: Exploring Their Role in Education*, Harvard professor John Miller points out that because the personal dimension of love is to be avoided in the classroom, the other love dimensions, of learning, compassion, beauty, and self-love are often given short shrift or completely ignored. For fear to be overruled by love, the many forms of love will need to be universally taught and expressed. And the best teachers will be us, as humanist leaders and role models to those who follow us, work for us, live with us, and engage with us. As we seek to teach, we can also seek to learn from other humanist revolutionaries, past and present. We can intentionally recruit people to our cause and our organizations who exhibit the seven I's attributes. These are the people who already understand that marching towards greater human progress on the back of personal growth and behavioral change is the only way forward. In his book *Thank You for Being Late: An Optimist's Guide to Thriving in the Age of Accelerations*, author Thomas Friedman acknowledges the importance of seeking the seekers when he wrote,

> The principal factor promoting historically significant social change is contact with strangers possessing new and unfamiliar skills.

Seekers carry capacities that we do not have while sharing many of the same sensibilities. They are driven not by winning or self-interest but by the idea of helping others as they seek better solutions for society's problems. Seekers are insatiably curious and open to the perspectives of others as the best means to find a collectively workable answer. And while there is no priority in the list of the seven revolutionary attributes, inclusion as accessibility may have a slight edge. To create a world that puts our humanity first, we must be open to all.

As we find others who lead from the heart, we will also set intentions that reflect the answers to the question "What does it mean to be human?" intentions that set clear expectations and re-establish the bedrock human principles of truth, ethics, morality, and integrity. And technology has no say is this matter because it has no morality, no conviction, no purpose other than function. Which is why as the technology train has hurtled ahead, these principles appear to have been left on the side of the tracks. They have been forgotten or ignored by many, the consequence of both technology itself and the persistence of self-interest. Humans who text while driving choose their need for convenience or pleasure over the safety of the people in the cars around them. The same is true of those who were opposed to wearing COVID-19 face masks and those not willing to be vaccinated. They choose individual freedom over collective health and concern for others. The inability of America's two political parties to work together reveals that one or both sides choose their party and its biases more than they honor the task of solving our nation's problems together. Our unwillingness to pick up trash off the street, to help a homeless person, or to step in when somebody is doing wrong to someone else is all about being selfish or at least being unwilling to sacrifice. It appears as if more and more humans are, consciously or not, showing a lack integrity in our choices, choosing short-term gratification or personal gain over long-term collective benefit—the tragedies of the commons repeated. How is it possible to achieve greater human progress if we have lost the human capacities that separate us from all other species? If we have become no better than animals fighting for our share of the kill? As leaders of corporations, organizations, communities, and families we can work to re-set the equation, making sure that our code of conduct, of ethics, morality, and humanity is at the center of every decision and every action. The societal bias away from truth and honor, from we to me, can be offset by our behaviors as leaders with integrity. At the same time the re-make of the core systems we rely on must both lead from the heart and make it possible to lead from the heart. With renewed effort humanism can once again become the center point of our systems, their primary intention and ultimately our motivating incentive.

OUR COURAGE IS A CHOICE

For revolutionary leaders and their followers these choices are, at their most foundational, about choosing courage. It takes courage to recognize the dark side of pure self-interest and to trade our disposition towards excessive personal comfort for a willingness to care for others. It takes courage to acknowledge and accept that our instinctive behaviors are selfish and to change them. It takes courage to exhibit ethical and moral integrity and to always act in alignment with the truth that we are a mutually reliant species on one planet with finite and increasingly depleted resources. Courage is a choice, a re-occurring decision to be made that will be made more frequently by more people, not because we are fundamentally braver than other people but because we choose courage over fear as the right thing to do in order to do the right thing. Courage is the central choice of the Humanist Revolution and perhaps its most defining behavior and intention. One final word from Kornfield:

> Neuroscience has shown that human beings are born with innate compassion and care for self and others. It also shows that human beings are born with survival circuits, which, when activated, operate from fear, aggression, selfishness, and hate. It's up to us which one we let create our future. Intention is the key.

ADAPT OR DIE

Alongside the need for more human-centric intentionality by us as leaders, comes the importance of the second I, innovation as adaptation. As is perhaps now painfully clear, the unrelenting pace of change demands collective, perpetual adaptation; changing the ways of us demands all of us always being willing and able to change. Therefore, revolutionary leadership involves motivating, teaching, and enabling our fellow citizens, employees, family, friends, and ourselves to do so. Even though much of the developed world operates as if the way things are is the way they will always be, that instinct to hold on to the status quo is no longer an option. Andy Grove, Intel's co-founder stated it starkly:

> You have no choice but to operate in a world shaped by globalization and the information revolution. There are two options: adapt or die.

Our stubborn preference for the status quo is perpetuated by our vague human progress intentions and a Maslovian tendency to close our eyes and hold on tight until the storm is over. The challenge in this case is that the growing storm, the gap between technology and us, and the many existential threats, cannot be fully vanquished. Given humankind's general reluctance to change, Grove's adapt or die declaration may appear as a threat itself. It should be considered instead as a gift, a supportive call for each of us to step forward and embrace the opportunity to evolve ourselves and those around us, and in doing so realize more from our lives while we give back to a world so desperately in need.

As we seek to accelerate our adaptive capacities, we must first accept that externalities when met with the right or wrong behaviors are the greatest determinant of where all this ends up. And that deeper, more intimate, behavioral human understanding is therefore key, both the behaviors that are detrimental and the ones that lead us to the truth and the collective outcomes we hopefully seek. Like the intention to love, the quest to better understand the other, whether the other is a stranger, a customer, a partner, an employee, or oneself, is also fraught with moments of fear and even discomfort. As transformative leaders we must be open to transforming ourselves and to begin the uncomfortable task of looking in the mirror and asking hard questions including: What are my behaviors that motivate and support the change of others? What are my behaviors that get in the way? Does that send the wrong signal? If I wish the people who follow me, collaborate with me, or support me to become more open, courageous, loving, and transformative to realize our shared human intentions, how am I doing on a daily basis exhibiting those desired behaviors? Victor Frankl, the neurologist, philosopher, Holocaust survivor, and author of *Man's Search for Meaning* wrote,

> Between stimulus and response there is a space. In that space is our power to choose our response. In our response lies our growth and our freedom.

Ours is fundamentally a challenge of mindfulness, of recognition of our biases and predilections, and the willingness and freedom to choose differently.

The second step in building adaptive capacity is to categorize the externalities, from threats to opportunities, to determine what degree of influence they should have on our adaptation of new behaviors and the speed at which that adaptation must occur. The need for external perspective carries with it the responsibility to create an environment of active teaching and learning. By definition, the word adaptation means to change in response to circumstance, to both take on new behaviors and to learn how to do new things, to become something other than how you are. Most human collectives, from families to countries, are not committed learning organizations which means they will struggle with the adaptation task. As discussed above, much of the developed world thinks education, the formal evolution of one's intellectual self, ends at ages 18 or 22 when in fact, adaptation and learning are now lifelong constants. For revolution leaders the only way to lead the revolution is by teaching and learning, not of functional skills but of behavioral ones—how to help employees get more comfortable with change, how to be better at collaboration, how to hold each other accountable, how to appreciate constructive confrontation, how to engage with uncertainty, and yes, how to lead with love. Addressing the climate crisis will require all these capacities and behaviors, and a level of global collaboration not seen since World War II. Our poor collective performance during the pandemic suggests we have a lot of work to do.

CREATING CONNECTIVE TISSUE

Changing the ways of us includes re-defining and prioritizing the task of connection as a core behavior. Even with the clearest of intentions and the most adaptive capacities, without a full embrace of the importance of connectivity our efforts will fail. The transformative criterion of connectivity will manifest in multiple forms. As the leaders of the revolution, we should first focus on re-connecting to ourselves and to each other. In her book *Alone Together: Why We Expect More from Technology and Less from Each Other*, Sherry Turkle, the MIT Professor and founder director of the MIT Initiative on Technology and Self succinctly captured how technology has changed the nature and integrity of social connections:

> But when technology engineers intimacy, relationships can be reduced to mere connections. And then, easy connection becomes redefined as intimacy. Put otherwise, cyberintimacies slide into cybersolitudes. And with constant connection comes new anxieties of disconnection.

Turkle goes on to explore the changes in our context of living by comparing the lens on life of the 19th century American naturalist and philosopher Henry David Thoreau versus our modern lives lived through screens, a constant modality likely resulting in more human disconnects:

When Thoreau considered "where I live and what I live for," he tied together location and values. Where we live doesn't just change how we live; it informs who we become. Most recently, technology promises us lives on the screen. What values, Thoreau would ask, follow from this new location? Immersed in simulation, where do we live, and what do we live for?

It's hard to imagine how we can achieve material gains in human progress if more and more humans are increasingly feeling disconnected, and, because of that, anxious, depressed, and lonely. As leaders we can wrest the ubiquitous control and context of technology away from our addicted hands and replace it with the context of being human. It's time that we actively, intentionally stewarded technology through the lens of our human needs and strengths, not our vulnerabilities and frailties. It is time we find our way back to a healthy intimacy that is founded on love, understanding, patience, and compassion, not on the false gods of transaction frequency and self-validation. It is time we actively work to replace the technology-accelerated anonymity of the neighborhood and the workplace with an effort to introduce us to us and the other while encouraging an attachment to both purpose and place. We can slow down and let go of both volume and pace to replace them with understanding. As Jerome Segal wrote in his book *Graceful Simplicity*:

> The time we give to things reflects our values. When everything is rushed, then everything has been devalued.

Greater human connectivity and with it a slowed, more natural transfer of meaning and value will result in more humane and effective organizations and communities and greater gains in human progress on every scale.

LEARNING FROM NATURE

Turkle's earlier reference to Thoreau points to another form of connection that the Humanist Revolution must explicitly re-establish and that is our respectful and loving connection with nature and our planet. As the COVID-19 pandemic exposed a multitude of difficult truths about the human condition and modern society, it also served as a painful reminder that our ongoing abuse of nature has life-threatening consequences. Human nature is at war with Mother Nature and technology has been the nuclear armament. It has been theorized that the coronavirus itself was enabled by the unnatural bridging between ecosystems and species and humankind's continued encroachment into wildlife habitats. We should stop our blindly destructive behaviors and willful ignorance of the impact of annihilating species and resources that make up the finite system called earth, the ripple effects of which are mostly unseen and yet incalculable. The revolution will work to reset the relationship between us and the planet and require all of us to learn to honor what Alexander von Humboldt, the world's first true naturalist saw as one planet, one life form, where every variable was both cause and effect to every other variable, including humankind. In 1800, Humboldt summarily declared that the fate of us was tied to the fate of our planet and vice versa. Two hundred and twenty years later we are just now realizing how correct he was.

As we acknowledge Humboldt's foresight, we should delve further into the human-forged glacial age known as the Anthropocene (following the current Holocene period). A term coined in 2000 by the Nobel Prize-winning chemist Paul Crutzen and biologist Eugene Stormer, their much-debated Anthropocene label was meant to capture this uncomfortable phase in history when the ways of us, our disconnected choices, and behaviors, really started taking a toll on the ways of it (Earth). In his 2019 article for *The Atlantic* titled "The Cataclysmic Break That (Maybe) Occurred in 1950" Robinson Meyer called out his conversation with Jan Zalasiewicz, a professor of geography at the University of Leicester and a member of the Anthropocene Working Group, a collection of academics and researchers charged with deciding what exactly to do with the Anthropocene label and if deemed true, establish when it started. Meyer's capture of Zalasiewicz's commentary suggests that the start date coincided with the exponential advances in technology and by extension industrialization that began around 1950:

> "If you look at the main parameters of the Earth-system metab-olism, then … things only began to change sharply and dra-matically with industrialization," he told me. He believes that the most significant event in humanity's life on the planet is the Great Acceleration, the period of rapid global industrialization that followed the Second World War. As factories and cars spread across the planet, as the United States and U.S.S.R. prepared for the Cold War, carbon pollution soared. So too did methane pollution,

the number of extinctions and invasive species, the degree of surface-level radiation, the quantity of plastic in the ocean, and the amount of rock and soil moved around the planet.

In seeking to change the ways of us, it would serve to follow nature's lead versus take advantage of her. The planetary abuse of the Anthropocene should end as quickly as it began, to be replaced by an age of connectivity and restoration, an age where all of nature, human and not, works in symbiotic, now revolutionary ways.

The Humanist Revolution's necessary connection with nature is both a corrective task and a restorative one, helping nurse nature back to health while tapping into her as a therapeutic balm for our own mental, emotional, and physical needs. During the pandemic millions of quarantined people sought the healing power of the outdoors. National parks in America were over-run, campgrounds had waiting lists, custom-fitted camper vans were on back order. People intuitively knew that connecting more with Mother Nature (and less with technology) would help reduce their stress and feel better, an intuition that is increasingly supported by science, as confirmed by Lisa Nisbet, a PhD in psychology at Trent University:

> There is mounting evidence, from dozens and dozens of researchers, that nature has benefits for both physical and psychological human wellbeing...You can boost your mood just by walking in nature, even in urban nature. And the sense of connection you have with the natural world seems to contribute to happiness even when you're not physically immersed in nature.

More recently, the author and MIT professor Alan Lightman wrote a moving article for *The Atlantic* titled "This is No Way to Be Human", sharing this,

> ...there is a profound disconnect between the natureless environment we have created and the "natural" affectations of our minds. In effect, we live in two worlds: a world in close contact with nature, buried deep in our ancestral brains, and a natureless world of the digital screen and constructed environment, fashioned from our technology and intellectual achievements. We are at war with our ancestral selves. The cost of this war is only now becoming apparent.

By connecting more with nature, we will no doubt discover other lessons and motivations, essential truths that will serve as guiding lights for our good behaviors and the work we need to do to realize our human progress intentions. Nature is an exemplar of adaptive capacity, an idea that is the center point of a field called biomimicry, first introduced by an American biophysicist named Otto Schmitt in the 1950s. Schmitt saw that the hyper-adaptive systems of the natural world could be used as models and guides for how to build synthetic and mechanical systems to solve complex human problems. The first plane invented by the Wright brothers in the early 20th century followed the aeronautical form of a pigeon; synthetic versions of shark skin are used as impermeable industrial coatings; the biomechanics of geckos inspire entirely different climbing systems. The perfected systems of nature are increasingly being used to solve functional problems while also serving as a benchmark for aesthetics and form. In his work Schmitt realized that the nature of nature is without bias to external information, but rather completely and objectively reliant on it as a feeder to the development of its own abilities to continue to survive. In natural systems there are no unhealthy behaviors to correct, no preconceived notions to disavow. The food chain is brilliantly simple in its construct, with every organism reliant on the other organisms on either side of its position in the chain both to survive and to motivate its adaptation. The systems of nature are instructive in their capacity to evolve separately and together, and we should connect to and learn from that.

That symbiotic systems view of nature extends to the question of how we as Humanist Revolutionaries can begin to imagine a different doctrine to craft as we seek to solve the problems of our world. Nature is the ultimate role model, providing an explicit blueprint for that doctrine, a sort of post-revolutionary sustainable operating plan. In her 1996 book *Biomimicry: Innovation Inspired by Nature*, biologist Janine Benyus explained,

> After decades of faithful study, ecologists have begun to fathom hidden likenesses among many interwoven systems. ...a canon of nature's laws, strategies, and principles...
> Nature runs on sunlight.
> Nature uses only the energy it needs.
> Nature fits form to function.
> Nature recycles everything.
> Nature rewards cooperation.

Nature banks on diversity.
Nature demands local expertise.
Nature curbs excesses from within.
Nature taps the power of limits.

I ask that you re-read Benyus's list, replacing the word "Nature" with the word "Humanity." In that light we can see nature as the ultimate revolutionary leader, exhibiting clear intentionality, profoundly adaptive capacity, intricate, sensitive, and highly efficient connective design, and the fourth leadership "I" criteria: inclusion known as accessibility. The symbiosis and interplays of her systems and sub-systems know no bounds or boundaries. While ecosystems exist, they operate as part of one great ecosystem, with the mutual reliance extending from the smallest of living things to the largest. We, like nature, can work to establish that same "oneness" by taking down the walls between us and between us and nature. We can change the ways of us by eliminating the belief that all subjects, components, functions, and entities are discrete, when in fact they are part of a mutually reliant whole. We can walk away from the legacy notions of cities, countries, and continents, of borders and protocols that theoretically separate us. We can soundly reject the tendency to think that if others don't look like us there is no place for them in our world. The nationalist and anti-globalist movements are walled propositions whose ideologies are antithetical to the need for inclusion and the now embedded connectedness we share. They are also woefully ignorant of its benefits and the associated requirement of constant collective adaptation. When queried regarding his stated support for globalization versus the anti-globalists, the Dalai Lama, Tibet's spiritual leader responded,

> The growing opposition to globalization is dependent on our reluctance to accept the principle that everything is perishable: meaning, the fact that everything is subject to permanent change. For Buddhists this is one of the basic pieces of wisdom that one must accept.

Nature accepts change and embraces connectivity and accessibility as its lifeblood. We can too. And we can stop behaving as if life is a zero-sum game that each of us seeks to win while ensuring that the other loses. The legacy walls of geography and hierarchy have no place in a world that is globally challenged. The idea that globalization is a choice is folly. We have one planet and one chance to change our behaviors to insure Earth's survival and ours.

THE TIME HORIZON IS NOT OURS

Like the prescription for the makeover of our society's core systems, we should also re-frame the time horizon for the Humanist Revolution's success, for the measured consequences of our newfound behaviors as its leaders, and the incentives that motivate those behaviors. Perhaps one of the most daunting tasks is that changing the ways of us requires accepting that it's not actually about us, or more pointedly, that the impact of our actions will likely not be realized in our lifetime. As Mahatma Gandhi once wrote,

> It's the action, not the fruit of the action, that's important. You have to do the right thing. It may not be in your power, may not be in your time, that there'll be any fruit. But that doesn't mean you stop doing the right thing. You may never know what results come from your action. But if you do nothing, there will be no result.

The time horizon of the revolution is from now until forever or at least the very distant future.

THE MATTER OF INCENTIVES

The sixth "I", Incentives, may be the most pivotal, because we humans don't tend to change unless we really must (we're desperate) or we really want to, when there is something explicitly in it for us. That thing is called an incentive. Accepting an infinite time horizon means walking away from the incentive of self-interest, the interests of our children and grandchildren, and the interests of our homelands, replacing it with the incentive of the common good. It requires turning the tragedy of the commons upside down or perhaps inside out. The cold qualifier, of course, is that history and our baked-in biases are against us being willing or able to do this. Despite broad agreement that loving one's neighbor and leading with openness and love are desirable traits, despite centuries of religious traditions espousing love and selflessness as the keys to the kingdom of heaven, and despite our personal desire to do right by each other, most of us tend to read the call for humanly love as completely unrealistic and impossible. That's just not the way the world (or we) works, we might lament, with good reason. Humans oppress, scam, harm, and even kill each other with distressing frequency, out of fear, greed, or primal force. Despite our common good aspirations, it does appear that the incentives we have built into our core human systems push us away from love and toward competition, self-maximization, dominance, and even oppression, all fueled by self-interest. We have come to accept, through our own fabrication and generational repetition, that this is "just human nature." We behave the (selfish) way we do because of the way we are and believe we need to be in order to survive. We're back to Maslow.

SURVIVAL AND SELF-INTEREST

It is true, at some level. All living things are required to compete for survival. That is a fact of life, of earthly nature because ours has always been a world of scarcity in some form. Ease and guaranteed comfort are rare in life and are quickly frittered away when they do by chance arise. It is perfectly reasonable, from an evolutionary perspective, that we should be competitive with and even, when required, exploitative of each other. So, we come by our orientation towards self-interest naturally, a trendline that has only been accelerated and magnified by two accidental developments: capitalism and technology. Since the dawn of the Enlightenment and the embrace of the invisible (demand/supply) hand, self-interest has become a sort of cultural imperative. Eighteenth century economists and philosophers including Locke, Hobbes, and Smith postulated the forerunning fundamentals of market-based capitalism, an integrated system of individual rights, egocentric motivations, and "invisible hand" economics that allowed for individuals to act in their own self-interest in competitive environments that eventually steered their actions toward the public good. This system of self-interest scaled very effectively with minimal design and delivered prosperity at a level never before seen. Technology made it go exponentially faster. The world has never known the economic capacity that we have today. For the first time in our history the majority of humanity lives far above survival and subsistence level and many of us enjoy comfort and luxury far in excess of that baseline. Self-interest works, sort of, until the unintended consequences pile up to the point that the system comes crashing down.

But that does not need to be the end of our story. Yuval Harari, in *Sapiens*, places our transition as humans from "animals of little significance" to a juggernaut that came to dominate the planet to the point at which we learned to cooperate, share knowledge, and build collective advantage. It seems just as possible that collaboration, rather than competition, may be the "secret sauce" of our future human success. It is certainly the path of the revolution.

Harari and others have noted that from the earliest tribal groups, this instinct for cooperation might have been limited to groups of 150 or so individuals but that over time we learned how to collaborate across larger and larger numbers. That intentional collaboration, the requisite connectivity and shared sense of responsibility for the whole, as both incentive and intention, have been present before, which means that they can be present again. In their convention challenging book, *The Dawn of Everything: A New History of Humanity*, authors David Graeber and David Wengrow explore whether self-interest has always been the dominant incentive of our existence, and suggest that in fact, it has not. They reveal numerous examples of civilizations that did not succumb to the self-interest game, suggesting that more egalitarian models of governance and existence are not beyond our reach. In an excerpt from their book, they too seem to be calling for the Humanist Revolution:

> What we need today is another urban revolution to create more just and sustainable ways of living. The technology to support less centralized and greener urban environments—appropriate to modern demographic realities—already exists...In the face of inequality and climate catastrophe, they offer the only viable future for the world's cities, and so for our planet. All we are lacking now is the political imagination to make it happen. But as history teaches us, the brave new world we seek to create has existed before, and could exist again.

Even with the comfort of the authors' interpretation of the past, we are still left with the nagging question of incentives, and more specifically, motivation. What influence does each of us need to pick up the flag and the shovel of the Humanist Revolution? To embrace the importance of the collective good, of human progress over the importance of ourselves? To sacrifice in the fight against our existential threats? To gain control of the technology train? What force could possibly compete with the force of self-interest, of personal gain? I believe the answer lies in our morality and its capacity to persevere, as reflected in a 2020 article by Dr. Ralph Lewis for *Psychology Today* titled, "Where Does Morality Come From?" He writes,

> Morality can be understood as a natural characteristic emerging from traits that are based on instinct and emotion and shaped by reason, evolving in an intensely social context...Cooperation and compassion have taken an especially strong lead over violence and cruelty in modern educated secular democratic societies, when contrasted with most of human history...Societal progress in our modern era has been uneven and faltering; catastrophic derailments have occurred along the way and will always be a risk. But the long-term positive trend toward more compassionate, cooperative societies has been strong and unmistakable. Societal ethics and compassion are achieved solely through human action—we have only ourselves and our fellow human beings to rely on in this collective human project.

Morality is what makes the singular incentive of self-interest distasteful, the "red in tooth and claw" nature of our systems alienating, and the growing inequities felt to be both unjust and inhuman. The incentive for us is fundamentally a moral one, to do right, because it is right. The growing concerns about the state of the world are fundamentally moral concerns. We just need to tap into that vein of simmering anger and dissatisfaction. It should be more than sufficient.

Whether you believe the Anthropocene began in the 1950s or not, we've been doing damage to the world and each other for thousands of years. It will take hundreds if not thousands more to get humankind and Earth back into balance. With this extended time horizon must come the acceptance of our mortality and the realization that the meaning of our life is not to live longer or amass more wealth but to leave the world a better place, to contribute to whole human progress. That is our summary incentive, a motivation of collective aspiration versus the force of individual desperation. Desperation is the feeling of fear that will arrive when the rising waters are at our door or perhaps the entire Internet has been hacked and all of us are being held hostage. We don't need to wait for that day. This moment in time, as we emerge from the pandemic, is an opportune time to change our behaviors, to change the game, as beautifully captured by Arundhati Roy in her book *The Ministry of Utmost Happiness*:

> Historically, pandemics have forced humans to break with the past and imagine their world anew. This one is no different. It is a portal, a gateway between one world and the next. We can choose to walk

through it, dragging the carcasses of our prejudice and hatred, our avarice, our data banks and dead ideas, our dead rivers and smoky skies behind us. Or we can walk through lightly, with little luggage, ready to imagine another world. And ready to fight for it.

Our imagined new world is a world of collective well-being. A world of infinite human latitude. A world of caring by the whole for the whole. And changing the ways of us is the only way to reach it. As daunting and even distasteful as the task may seem, we at least have a checklist of transformative criteria: intentionality, adaptability, connectivity, accessibility, expanded time horizons, and the right incentives. And to deliver against those criteria requires a seventh pursuit, and that is individual responsibility, or more colloquially, activism. As the author Sue Monk Kidd once expressed, "There's a gap somehow between empathy and activism." As revolutionaries it is not enough to care, we must work to bridge that gap, to take the wheel of the train.

Ours must be a conscious commitment to changing our individual and collective ways, fueled with passion and intensity, a determination that *we* can replace *me* as our guiding force, and a belief that our divisions can be broached. This revolution maybe the ultimate selfless act, an intimate undertaking, and perhaps the hardest thing any of us will ever attempt to do. To transform the world, we will first need to transform ourselves. And we can. There is no natural law that dictates that we must be driven by self-interest. No principle of physics that the tragedy of the commons must persist. There are no laws that protect our legacy notions and systems. There is no rule that says to lead with love is weak. Fear is not an absolute because courage is a choice, as is having the integrity to always do the right thing.

A 21ST CENTURY HUMANIST CODE

As revolutionary leaders, our ability to achieve the seven "I's" success criteria of societal transformation is based on exhibiting some essential behaviors. In her book *Humanly Possible*, Sarah Bakewell captures the centuries-old history of humanist movements and the stories of the brave individuals who stepped forward to challenge the prevailing societal measures and means and to declare a far different, more human set of behavioral intentions. In her conclusion Bakewell references the Humanists International's 2022 "Declaration of Modern Humanism", a compelling

600-word manifesto that presents what are referred to as "essentials of the humanist worldview." To my mind our species needs an even more directive tool, a short, explicit code of the behaviors that we must all live by to make our revolutionary contribution. This Humanist Code is not just for us revolutionaries but for the 10 billion people that will ultimately inhabit our world. It is for everybody, from our political, corporate, and community leaders to the 250 babies born this very minute. The Code's seven declarations should be viewed as a draft, a starting point for a collective global conversation about the universal behaviors that will result in the outcomes we want and need.

(1) **We shall make collective well-being the primary measure of all progress.**
Let us make this shared human intention the purpose of every action we take, every system we build, every relationship we form—from how we lead, build, design, and work to how we parent, partner, participate, and live. Self-interest, avarice, gluttony, and the outcomes they produce have no role in our collective existence.

(2) **We shall care for all living things.**
There is only one world, one resource, one life to be shared, and cared for by all for all. Nature is our mother, all humans are our brothers and sisters, and all creatures are part of our family. We embrace the interdependence between us, between cultures and wealth, between colors and beliefs. Borders, boundaries, and walls are human fabrications built by fear and greed. Let us learn to love, not just tolerate, diversity. We will replace pride of self with pride of all. We honor life.

(3) **We shall seek the truth.**
Without truth we are lost. With truth we can realize the full potential of the human experiment. The new march to and measure of human progress requires that we find our way back to the foundational truths that we share. Our search must come with a willingness to let go of our current narrative and our inability to hear things we don't want to hear. Our need for certainty will be overruled as we allow the truth to set us free.

(4) **We shall live slower, go deeper, and unplug.**
Let us not fear doing less or slowing down. Let us become more selective with our attention and work harder to understand ourselves

and the world we live in. We will reduce our celebration of the trappings of modern life and replace our insatiable need for distraction, transactions, and convenience with quiet contemplation and a singular desire to foster our own well-being and that of all others. Our lust for things will be exchanged with a lust for life.

(5) We shall let go of ego, make sacrifices, give, and forgive.

Let us replace self-interest with self-sacrifice. Our life's meaning is to make that contribution manifest——to serve, to help, to give all that we can to advance our human cause and come to the aid of others. We will collaborate freely, share openly, compromise often, and bridge our divides by letting go of our biases and dislikes, offering in their stead forgiveness and gratitude. We will listen more than we speak. We will say and give thanks often. Our desperate search for validation is over.

(6) We shall love ourselves and each other.

Human progress requires far greater human connection, more intimacy and much less anonymity, beginning with us. Let us live by name, engage with all those we can see and hear, and say hello to strangers. Let us own our role in the world and our responsibility to be welcoming to all living things. Our ability to connect authentically and at depth is a defining capacity of humans and a central contributor to our well-being. To love another requires first loving ourselves. Envy has no place, sustained anger towards anyone has no reason.

(7) We shall lead.

Let us each become a humanist revolutionary. Let us each envision the world we wish to inhabit and then work to make it so, in whatever ways we are able, however small they may be. The butterfly effect applies in the affairs of both Mother Nature and human nature. Let us each step forward to demand the humanization of technology, to take the wheel of the technology train and steer it to its best, most human destination. Let us each push for systems change and new paradigms of rule singularly focused on our collective well-being. Let us each lead with love. Others will follow. There is no time to stand still.

A FINAL CALL TO ACTION

The summary question is not whether we can learn to steer the technology train and in doing so bend the arc towards a world that is fairer, kinder, and more focused on the well-being of our entire species and the planet we rely on, but who, who will lead the change?

You now know the answer is us. Margaret Mead, the American cultural anthropologist once wrote,

> Never doubt that a small group of thoughtful citizens can change the world. Indeed, it is the only thing that ever has.

What each of us chooses to do, and how we choose to do it, is really the most important means and measure, the essence of the revolution itself.

Act.

Epilogue

Re-writing Our Narrative

"Sometimes things become possible if we want them bad enough."

T.S. Eliot

Western society and the developed world in general have suffered from a problematic, unintended convergence over the last few hundred years. Exiting the Middle Ages and its defining pandemic conclusion, the societal pendulum slowly swung towards an adulation of rational thought, the sciences, and a growing conviction that human cognition sat above all else, not our spirit but our mastery of the physical world. The Renaissance, which begot the Enlightenment, was the beginning of the end of spiritual centricity to our lives and brought with it a sense of omnipotence that we could analyze all, understand all, and achieve mastery of ourselves and the world around us. We turned out to be wrong. And the rapid advance of technology and all its functional successes made our erroneous conceit even worse.

The swing toward reason and science, which fueled industrialization, brought with it the rapid rise of capitalism and the celebration of individual freedoms over collective, moral, and spiritual responsibilities. Slowly we began to lose touch with what it means to be human, to be alive, to be part of a greater whole. The concept of "we" was gradually overwhelmed by the dark seductions of "me." This constricted view of human purpose and of human behavior narrowed even further as the technology train accelerated away from us. It has resulted in the troubling gap, a growing roster of existential threats, but most importantly a void within us and between us— between the way we should live and the current, false narrative we share. That narrative is a collection of stories, nothing more or less than stories.

DOI: 10.1201/9781003089902-15

The systems the Western world has built and the modern behaviors we celebrate require us to deceive ourselves every minute of every day. Ours are implicit delusions supported by a longstanding, deeply ingrained narrative that mistakenly holds up economic progress and individual freedoms as the sole measures of a healthy society and the way we live. In our technology-fueled, self-interest-fixated, present-day narrative we talk about our global economy delivering prosperity to all those willing to work. This is manifestly not true. Millions still live in poverty despite their unrelenting toil.

We talk about businesses and institutions caring for and loving their employees and customers. And yet our systems reward the sale of unhealthy products, the abuse of labor, and the destruction of the planet. Global stock markets are unabashedly focused on growth and profit. The idea of corporate purpose has little value to most of today's investors. Competitive capitalism, fueled by self-interest, requires businesses to addict their customers, fire their employees, and ruin lives with impunity. It's just business, we explain to ourselves and our children.

We talk about politics and government working to build a strong and just society, yet the system is increasingly and blatantly about gaining power, manipulation, and control. And more and more we accept that our governing leaders do not have to pretend otherwise. The current narrative has been grossly bent to accommodate a loss of human integrity in our leaders' leadership and their abdication of whole human progress intent. We accept these outcomes because we convince ourselves that we are powerless to demand different ones.

We deny climate change is a crisis and even those of us who accept it do little to change our behaviors and help mitigate the threat. The disconnect between what we say and what we do is yet another fictional chapter in our modern-day story.

We talk about technology and progress making our lives better and more fulfilling, while progress, as we have simplistically conceived it, just delivers ever more and more material goods, distracting us from our uniquely human needs for self-knowledge, self-actualization, and connection with each other and the world around us. We have lost our sense of well-being.

The narrative much of the developed world holds up as the success story to be adopted by the rest of the world is, in actuality, a narrative of broken truths, subtle deceptions, unhealthy consequences, and human mistakes. We all know this in our hearts, yet we are reluctant to acknowledge it, perhaps in part because we feel complicit and because we have no concept of how to re-write it. In 2020, the Robert Woods Johnson Foundation published *Well-being: Expanding the Definition of Progress*. Its authors called out the role of narrative as both part of our problem and the solution:

> Historic practices, particularly in capitalist or industrialized cultures, have trained people to place personal gain ahead of societal well-being, and to put the wealth of their own state or nation before global equity and sustainability. Systems, structures, cultures, and dominant narratives reinforce and reward this mindset and the behaviors that follow. For a new narrative to gain currency, and for public demand and support for a well-being approach to increase, it is important to shift individual consciousness and individual definitions of progress. Shifting the focus from "me" to "we" is a profound undertaking, but the well-being narrative is a powerful tool for doing so. Building on shared values, the narrative connects people to each other, their environment, and the planet. The public will that results leads to greater expectations, demand and incentives for actions that support well-being.

What many of us in the West appear to have forgotten during the seismic changes that occurred beginning in the 20th century is that collaboration and community have always been an essential part of the human experiment— re-inforced through practical mutual reliance and an ordained expectation of public spirit. It is only in recent decades, thanks to the unchecked global juggernaut of transactional capitalism, the slow rejection of religion, and the steady erosion of person-to-person connection, that unbridled technology made the matter of the collective good and human responsibility seemingly moot.

Over 45 years ago, in his 1978 Harvard College address Alexandr Solzhenitsyn called out that flaw in our prevailing Western narrative, albeit in a starker way:

> "The West kept advancing socially in accordance with its proclaimed intentions, with the help of brilliant technological progress. All of a sudden it found itself in its present state of weakness…this means that the mistake must be at the root, at the very basis of human thinking in the past centuries…everything beyond physical well-being and accumulation of material goods, all other human requirements and characteristics of a subtler and higher nature,

were left outside the area of attention of state and social systems, as if human life did not have a superior sense. That (loss of spiritual sense) provided access for evil, of which in our days there is a free and constant flow."

Whether we accept Solzhenitsyn's characterization of the unhealthy influences that have become so pervasive as evil or not, we should recognize that they have been and continue to be ultimately destructive. They are influences that can only be offset by re-writing our narrative and re-defining what it means to be human in today's world.

How then do we craft a new narrative that restores intimate human understanding and shared well-being as both the means and the ends of our story? There is a way, it will just require a humanist revolution, a radical re-invention of the human measures we hold up and the human means we deploy to realize them. Our task is akin to that of weavers, intertwining the essential threads of our humanity to form the fabric that will tie us and all living things back together. The threads are both the essential behaviors and essential models like the seven I's applied to the systems that underpin how society operates. The tapestry woven will support and celebrate our ability to prevail over self-interest to achieve outcomes of well-being and better choices for all, to bend the arc towards justice, to re-define our future for ourselves and each other. As we weave we must also be willing to model the behaviors every day and persistently pass on the humanist torch to those we love and those we do not know. For even when the destination has been reached, the revolution will not be over. The future of humanity is an infinite journey. With our current cognitive tools, we can only approach, but will never arrive, at the end. Human knowledge, in words and actions, will always be incomplete, approximate, and fallible. We can only learn what works by trying. Build, test, learn, repeat, and teach. Because the narrative will need to be re-written again and our children and their children must be ready, willing, and able to re-write it.

Onward.

ACKNOWLEDGEMENTS

The journey of writing *Technology is Dead* was inspired and enabled by many people; more evidence that the only way forward is together. I am profoundly grateful for every contribution. Sop Mohanty provided the spark, inviting me to Singapore in 2018 to deliver a speech titled *Technology is Dead* to thousands of public and private sector executives and technologists about the growing human imperative. Early in 2019, while the Managing Director of the Harvard Innovation Labs, I retained Faisal Matalqa, a Harvard University research assistant, to help me frame the book's outline and arc. When the pandemic arrived in early 2020 the book was well underway, but the quarantine period brought with it a period of self-doubt. My editor at Taylor & Francis Randi Slack's persistent positive feedback and encouragement got me through. In 2021, as the book took shape, I realized that there was a dimension of the story that I was struggling to capture. Serendipitously I re-connected with David Boghossian, a mentor from my days at the Harvard Innovation Labs. David had just listened to my podcast Insert:Human and offered his assistance. He became and still is a critical thought partner and contributor of important ideas and compelling language. As I neared completion of the book in late 2023 my sister Melissa Colbert offered to help with a final edit and the onerous task of documenting all the references. Her meticulous work, her way with words, and grammatical rigor have been a huge gift.

Along the way there have been many others who have provided me with motivating impetus and keen insights that were instrumental to crafting the book's argument and the solutions it offers up. The long list includes all of my Insert:Human podcast guests:

Marc Baker, Melissa Yahia, Mark Silverman, Sarah Seegal, Tom Furber, Eric Weiner, Richard Barrett, Kay Van-Peterson, Rana El Kaliouby, Billie Rosoff, Charla Jones, Beth Babcock, Julie Matheson, Julian Treasure, Sherry Harris, David Grinspoon, Dr. Madhavi Venkatesan, Blake Ezra, Mark Edwards, Linda Hoffman, Erin Baker, Laura Drake, Mike Horne, Mike Hutchinson, Ed Schein, Peter Schein, Beatrice Coron, Jon Levy, Bryan Welch, Raya Bidshahri, Dané Johnson, Jessica Pachuta, Dror Yaron, Dr.Laurie Leshin, Steven Titus Smith, Dave McLaughlin, Dr. Richard Harris, and Bob Contri.

There are a handful of others who also played a pivotal role: my 100-year-old surrogate mother Billie Rosoff who defines what it means to be human; my wife Kate Gilbert who dutifully read draft after draft and always offered positive feedback; Lara Zimmerman who gently nudged me forward when I was considering writing the book; Spencer Glendon who is my role model for intellectual rigor and meaningful climate activism; and my dear friend now gone, Alan Lewis, who constantly reminded me that a life without risk taking is no life at all.

And finally, I want to acknowledge all the organizations and individuals who have worked or are working to advance humanism and the human progress agenda, or just trying harder to be better human beings. Whether they know it or not, they are humanist revolutionaries. And I am deeply grateful for that.

Thank you all.

Bibliography

INTRODUCTION

Christia Spears Brown, October 26, 2023. "Here's Why States Are Suing Meta for Hurting Teens with Facebook and Instagram." *Scientific American.* www.scientificamerican.com/article/heres-why-states-are-suing-meta-for-hurting-teens-with-facebook-and-instagram/

Ikeda, Daisaku. 1973. *The Human Revolution.* Weatherhill.

LaFrance, Adrienne, June 5, 2023. "The Coming Humanist Renaissance." *The Atlantic.*

Wright, Lawrence. July 13, 2020. "How Pandemics Wreak Havoc—and Open Minds." *The New Yorker.* www.newyorker.com/magazine/2020/07/20/how-pandemics-wreak-havoc-and-open-minds

CHAPTER 1

Bellah, Robert, et al. 1991. *The Good Society.* New York: Vintage.

Branson, Richard, Quoted by Eleanor Lawrie, "'Our obsession with ownership is at a tipping point': As he gives £1m to UK start-ups Richard Branson reveals why he's backing the sharing economy:, *This Is Money,* www.thisismoney.co.uk/money/smallbusiness/article-3663869/Richard-Branson-obsession-ownership-tipping-point.html

Centers for Disease Control, August 31, 2022. "Life Expectancy in the U.S. Dropped for the Second Year in a Row in 2021."

Databridge Market Research, October 2022. "Global Influencer Marketing Platform Market – Industry Trends and Forecasts to 2029".

"Is Disintermediation the Future of Finance." *International Banker,* October 7, 2020. https://internationalbanker.com/finance/is-disintermediation-the-future-of-finance/

Minor, Lloyd. 2019. "A look at how data is democratizing health care." *Scopeblog.* Stanford University. https://scopeblog.stanford.edu/2019/01/22/a-look-at-how-data-is-democratizing-health-care/

MIT Technology Review. "10 Breakthrough Technologies 2023". February 26, 2020. www.technologyreview.com/10-breakthrough-technologies/2020/

Osztovits, Adam, Koszegi, Arpad, Nagy and Bence, Damjanovics. 2015. "Sharing or pairing? Growth of the sharing economy." *PwC.* www.pwc.com/hu/en/kiadvanyok/assets/pdf/sharing-economy-en.pdf

Pinker, Steven. 2019. *Enlightenment Now: The Case for Reason, Science, Humanism, and Progress.* London: Allen Lane.

Santos, Henri, Varnum, Michael E. W., and Grossman, Igor, July 13 2017. "Global Increases in Individualism." *Psychological Science.* https://journals.sagepub.com/doi/10.1177/0956797617700622

Schoch, Marta and Lakner, Christoph. November 5, 2020. *World Bank Data Blog.* https://blogs.worldbank.org/opendata/global-poverty-reduction-slowing-regional-trends-help-understanding-why

Toffler, Alvin. 1984. *Future Shock.* New York: Bantam.

2001. *Understanding the Digital Divide.* OECD Publications. www.oecd.org/sti/1888451.pdf

Wu, Tim. February 16, 2018. "The Tyranny of Convenience." *The New York Times.* www.nytimes.com/2018/02/16/opinion/sunday/tyranny-convenience.html

CHAPTER 2

Altman, Sam, March 16, 2023. "Interview by Rebecca Jarvis on ABC News."

Asimov, Isaac. 1988. *Isaac Asimov's Book of Science and Nature Quotations.*

Bellah, Robert, et al. 1991. *The Good Society.* New York: Vintage.

Carter, Ash. 2019. Keynote address at MIT *Conference on Technology Innovation and Public Purpose.*

Cook, Tim. 2018. Address at the *40th International Conference of Data Protection and Privacy.* Brussels. www.americanrhetoric.com/speeches/timcookeuprivacy.htm

Friedman, Thomas. 2016. *Thank You for Being Late: An Optimist's Guide to Thriving in the Age of Accelerations.* New York: Farrar, Straus and Giroux.

Harari, Yuval. February 7, 2015. "Interview with Arun Rath on NPR's All Things Considered." www.npr.org/2015/02/07/383276672/from-hunter-gatherers-to-space-explorers-a-70-000-year-story

Harari, Yuval Noah. 2017. *Homo Deus, A Brief History of Tomorrow*. New York: Harper Collins.

Marshall, Jenna. August 1, 2018. "The Growing Gap Between Physical and Social Technologies." *Santa Fe Institute*. https://phys.org/news/2018-08-gap-physical-social-technologies.html

Rudin, Cynthia, and Radin, Joanna, November 22, 2019. "Why Are We Using Black Box Models in AI When We Don't Need To? A Lesson from an Explainable AI Competition." *Harvard Data Science Review*. Issue 1.2, Fall 2019. https://hdsr.mitpress.mit.edu/pub/f9kuryi8/release/6?readingCollection=af83430a

CHAPTER 3

Bellah, Robert N. et al. 1991. *The Good Society*. New York: Knopf.

Einstein, Albert. 2000. *The Expanded Quotable Einstein*. Princeton University Press.

Havel, Vaclav. 1990. "1990 New Year's Speech." www.vhlf.org/havel-quotes/1990-new-years-speech/

Kahneman, Daniel. 2011. *Thinking Fast and Slow*. New York: Farrar, Straus and Giroux.

King, Dr. Martin Luther Jr. 1958. *Stride Toward Freedom: The Montgomery Story*. New York: Harper Bros.

Mike Allen, November 9, 2017. Parker, Sean quoted in "Sean Parker unloads on Facebook: 'God only knows what it's doing to our children's brains'". *Axios*. www.axios.com/sean-parker-unloads-on-facebook-2508036343.html?utm_medium=linkshare&utm_campaign=organic

Thoreau, Henry David. 1854. *Walden*.

CHAPTER 4

Aaron Sekhri, January 18, 2013. Ban, Ki-Moon quoted in "UN Secretary General Ban Ki-Moon addresses Mali, Syria, women's rights at Stanford." *The Stanford Daily*. https://stanforddaily.com/2013/01/18/ban-urges-international-cooperation-to-solve-global-challenges/

Applin, S.A., June 14, 2019. "Everyone's Talking About Ethics in AI. Here's What They're Missing." *Fast Company*. www.fastcompany. com/90356295/the-rush-toward-ethical-ai-is-leaving-many-of-us-behind

July 31, 2020. "Antibiotic Resistance." World Health Organization. www. who.int/news-room/fact-sheets/detail/antibiotic-resistance

June 17, 2019. "9.7 Billion on Earth by 2050, But Growth Rate Slowing Says New UN Population Report." *UN News*. https://news.un.org/ en/story/2019/06/1040621

Auxier, Brooke et al., November 15, 2019. "Americans and Privacy: Concerned, Confused and Feeling Lack of Control Over Their Personal Information." Pew Research Center. www. pewresearch.org/internet/2019/11/15/americans-and-privacy-concerned-confused-and-feeling-lack-of-control-over-their-personal-information/

Dave Davies on NPR Fresh Air, April 29, 2020. McNeil, Donald G. quoted in "Compared with China, U.S. Stay-At-Home Has Been 'Giant Garden Party.' Journalist Says." *NPR*. www.npr.org/sections/ health-shots/2020/04/29/847755751/compared-to-china-u-s-stay-at-home-has-been-a-giant-garden-party-journalist-says

October 30, 2023. "FACT SHEET: President Biden Issues Executive Order on Safe, Secure, and Trustworthy Artificial Intelligence." www.whitehouse.gov/briefing-room/statements-releases/2023/10/ 30/fact-sheet-president-biden-issues-executive-order

Gallagher, James, July 15, 2020. "Fertility Rate: 'Jaw-dropping' Global Crash in Children Being Born." *BBC News*. www.bbc.com/news/ health-53409521

Grissom, Adam R. et al. 2020. "COVID-19 Air Traffic Visualization." Rand Corporation. www.rand.org/pubs/research_reports/RRA248-7.html

Guterres, Antonio, December 2, 2020. "Secretary-General's address at Columbia University: 'The State of the Planet'" www.un.org/sg/en/ content/sg/speeches/2020-12-02/address-columbia-university-the-state-of-the-planet

Hedges, Chris, July 6, 2003. "What Every Person Should Know About War." *The New York Times*. www.nytimes.com/2003/07/06/books/ chapters/what-every-person-should-know-about-war.html

Homans, Charles, February 20, 2020. "A Disaster Video That Finally Tells the Truth About Climate Change." *The New York Times*. www. nytimes.com/2020/02/20/magazine/australia-fires-video.html

Klein. Ezra, June 5. 2014. "7 Reasons America Will Fail on Climate Change." *Vox*. www.vox.com/2014/6/5/5779040/7-reasons-America-fail-global-warming

Lebow, Richard Ned. 2010. *Why Nations Fight: Past and Future Motives for War*. Cambridge: Cambridge University Press.

2019. "Life Expectancy at Birth for Both Sexes combined (years)." United Nations Population Division. https://data.un.org/Data.aspx?d=Pop Div&f=variableID%3a68

Manjunath, B.S. "Covid-19: 8 Ways in Which Technology Helps Pandemic Management." *Economic Times*. https://cio.economictimes.indiati mes.com/news/next-gen-technologies/covid-19-8-ways-in-which-technology-helps-pandemic-management/75139759

May 2019. "Media Release: Nature's Dangerous Decline 'Unprecedented'; Species Extinction Rates 'Accelerating.'" Intergovernmental Science-Policy Platform on Biodiversity and Ecosystem Services (IPBES). www.ipbes.net/news/Media-Release-Global-Assessment

Ord, Toby. 2020. *The Precipice: Existential Risk and the Future of Humanity*. New York: Hachette.

Storr, Will. 2021. *The Status Game: On Social Position and How We Use It*. New York: William Collins.

2019. *This Giant Beast That is the Global Economy*. TV Series. www.imdb.com/title/tt9617860/plotsummary

October 31, 2020. "Twin Crises, Twin Cures: Disease, Biodiversity and Climate Change." *The Economist*. https://espresso.economist.com/16e62507eba0d973dc7aa14aa3aa41ab

May 6, 2019. "UN Report: Nature's Dangerous Decline 'Unprecedented'; Species Extinction Rates 'Accelerating.'" UN Sustainable Development. www.un.org/sustainabledevelopment/blog/2019/05/nature-decline-unprecedented-report/

Watkins, Johnathan, Charles, Dominic Charles, and Merkl, Andreas. 2022. "The Social Cost of Plastic-Related Harms." The Minderoo Foundation. https://cdn.minderoo.org/content/uploads/2022/10/13131230/The-Price-of-Plastic-Pollution-Annex-1.pdf

"Worldwide Spread of COVID-19 Accelerated Starting on February 19, 2020." Rand Corporation. www.rand.org/pubs/research_reports/RRA248-6.html

Wray, Christopher, October 30, 2019. "Global Terrorism: Threats to the Homeland." FBI News. www.fbi.gov/news/testimony/global-terrorism-threats-to-the-homeland-103019

January 15, 2020. "The Global Risks Report 2020." World Economic Forum. www.weforum.org/reports/the-global-risks-report-2020

CHAPTER 5

Alba, Davey, March 29, 2020. "How Russia's Troll Farm is Changing Tactics Before the Fall Election." *The New York Times*. www.nytimes.com/2020/03/29/technology/russia-troll-farm-election.html

Alesina, Alberto and La Ferrara, Eliana, September 2005. "Ethnic Diversity and Economic Performance." *Journal of Economic Literature*. https://scholar.harvard.edu/files/alesina/files/ethnic_diversity_and_economic_performance.pdf

Bell, K., October 23, 2021. "Facebook Researchers Were Warning About its Recommendations Fueling QAnon in 2019." *Engadget*. www.engadget.com/facebook-carols-journey-qanon-reports-005159230.html

Bellah, Robert N. et al. 1991. *The Good Society*. New York: Knopf.

Chakravorti, Bhaskar, July 16, 2017. "Is America's Digital Leadership on the Wane?" *The Conversation*. https://theconversation.com/is-americas-digital-leadership-on-the-wane-80936

Dalai Lama XIV. 2015. *An Open Heart: Practicing Compassion in Everyday Life*. New York: Little, Brown and Co.

2021. "DFA: Distributional Financial Accounts". *U.S. Federal Reserve*. www.federalreserve.gov/releases/z1/dataviz/dfa/distribute/chart/

Hammurabi. www.goodreads.com/author/quotes/206823.Hammurabi#:~:text=%E2%80%9CTo%20bring%20about%20the%20rule%20of%20righteousness%20in,that%20the%20strong%20should%20not%20harm%20the%20weak.%E2%80%9D

Garrison, David W., June 24, 2019. "Most Mergers Fail Because People Aren't Boxes." *Forbes*. www.forbes.com/sites/forbescoachescouncil/2019/06/24/most-mergers-fail-because-people-arent-boxes/?sh=7bc1cd3c5277

Glubb, Sir John Bagot. 1978. *The Fate of Empires and Search for Survival*. London: William Blackwood and Sons. Ltd.

Kang, Jaewon and Gasparro, Annie, March 15, 2020. "Grocers Fail to Keep Up With Demand as Coronavirus Pandemic Spreads." *Dow*

Jones News. https://uk.advfn.com/stock-market/EURONEXT/AD/share-news/Grocers-Fail-to-Keep-Up-With-Demand-as-Coronavirus/81993240

Marantz, Andrew, October 12, 2020. "Why Facebook Can't Fix Itself." *The New Yorker.* www.newyorker.com/magazine/2020/10/19/why-facebook-cant-fix-itself

Owens, Caitlin, July 22, 2020. "Parents Turn to "Pods" as a Schooling Solution." *Axios.* www.axios.com/parents-schools-coronavirus-pods-a18f0916-7dcc-43ff-bffe-5c33c753a23a.html

Peters, Jeremy W, October 7, 2020. Quoting Yuval Levin in "Mike Pence Brought Conservatives Home. What if They Don't Need Him Anymore?" *The New York Times.* www.nytimes.com/2020/10/07/us/politics/mike-pence-debate.html

Pope Francis, October 3, 2020. Encyclical "On Fraternity and Social Friendship." *The Vatican.* www.vatican.va/content/francesco/en/encyclicals/documents/papa-francesco_20201003_enciclica-fratelli-tutti.html

Pope Francis, November 24, 2019. "Address of the Holy Father on Nuclear Weapons." *The Vatican.* www.vatican.va/content/francesco/en/speeches/2019/november/documents/papa-francesco_20191124_messaggio-arminucleari-nagasaki.html

Storr, Will. 2019. *The Status Game: On Social Position and How We Use It.* New York: William Collins.

Williamson, Kevin D., August 19, 2014. "Homogeneity is Their Strength." *National Review.* www.nationalreview.com/2014/08/homogeneity-their-strength-kevin-d-williamson/

2017. "Bridges." *American Society of Civil Engineers.* www.infrastructurereportcard.org/wp-content/uploads/2017/01/Bridges-Final.pdf

CHAPTER 6

Babic, Milan et al., July 19, 2018. "Who is More Powerful—States or Corporations?" *The Conversation.* https://theconversation.com/who-is-more-powerful-states-or-corporations-99616

Brands, Hal and Sullivan, Jake, May 22, 2020. "China Has Two Paths to Global Domination." *Foreign Policy.* https://foreignpolicy.com/2020/05/22/china-superpower-two-paths-global-domination-cold-war/

Briner, Raphael, February 4, 2015. "The New 2015 Top 500 NGOs is Out." *The Global Journal.* www.theglobaljournal.net/article/view/1171/

Charlton, Emma. October 1, 2019. "6 Things to Know About China's Historic Rise." *World Economic Forum.* www.weforum.org/agenda/2019/10/china-economy-anniversary/

Chomsky, Noam. 2004. *Language and Politics.* AK Press.

Clancy, Tom interviewed December 6, 1995 by Sarah Schafer in "Vonnegut and Clancy on Technology." *Inc. Online Business.* www.inc.com/magazine/19951215/2653.html

"Democracy Index 2022." *Economist Intelligence Unit.* www.eiu.com/n/campaigns/democracy-index-2022/

Drake, Laura, February 19, 2020. "Book Review: Re-Engineering Humanity." *The Technoskeptic.* https://thetechnoskeptic.com/reengineering-humanity-review/

2020. "Global Risks 2020: An Unsettled World." *World Economic Forum.* www.weforum.org/publications/the-global-risks-report-2020/Global

Heimans, Jeremy and Timms, Henry. 2018. *New Power: How Power Works in Our Hyperconnected World—and How to Make It Work for You.* New York: Doubleday.

Kendall-Taylor et al., March/April, 2020. "The Digital Dictators." *Foreign Affairs.* www.foreignaffairs.com/articles/china/2020-02-06/digital-dictators

Manjoo, Farhad interviewed October 26, 2017 by Terry Gross of NPR's Fresh Air. "How 5 Tech Giants Have Become More Like Governments Than Companies." *National Public Radio.* www.npr.org/2017/10/26/560136311/how-5-tech-giants-have-become-more-like-governments-than-companies

Myers, Steven Lee, March 29, 2021. "An Alliance of Autocracies? China Wants to Lead a New World Order." *The New York Times.* www.nytimes.com/2021/03/29/world/asia/china-us-russia.html

Nazifa Alizada et al. 2021. "Autocratization Turns Viral." *Democracy Report 2021.* University of Gothenburg, V-Dem Institute. https://www.v-dem.net/democracy_reports.html

Nye, Joseph S. 1991. Bound to Lead: *The Changing Nature of American Power.* Basic Books.

Nye, Joseph S., June 24, 2004. "The Rising Power of NGO's." *Project Syndicate.* www.project-syndicate.org/commentary/the-rising-power-of-ngo-s-2004-06

Olsen, Sam, October 4, 2020. "China is Winning the War for Global Tech Dominance." *The Hill.* https://thehill.com/opinion/technology/518773-china-is-winning-the-war-for-global-tech-dominance

Tolstoy, Leo. 1869. *War and Peace.*

Turchin, Peter and Gavrilets, Sergey, September 2009. "Evolution of Complex Hierarchical Societies." *Social Evolution and History,* 8:2. 'Uchitel' Publishing House. http://peterturchin.com/PDF/Hierarch.pdf

Paris, Roland, December 2015. "Global Governance and Power Politics: Back to Basics." *Ethics & International Affairs.* www.ethicsandinternationalaffairs.org/2015/global-governance-power-politics-back-basics/#:~:text=No%20amount%20of%20institutio nal%20proliferation%20or%20innovation%20can,of%20politi cal%20legitimacy%2C%20war%20and%20peace%2C%20and%20c ommerce.

"World Economic League Table 2021." *Center for Economics and Business Research (CEBR).* https://cebr.com/wp-content/uploads/2020/12/WELT-2021-final-23.12.pdf

CHAPTER 7

Barrett, Richard. 2020. *Worldview Dynamics and the Well-being of Nations.* Lulu Publishing Services.

Coccia M. and Bellitto M. 2018. "A Critique of Human Progress: A New Definition and Inconsistencies in Society", *Quaderni IRCrES-CNR,* 4(3), 51–67. http://dx.doi.org/10.23760/2499-6661.2018.017

Easterlin, R. 1974. "Does Economic Growth Improve the Human Lot? Some Empirical Evidence" In: David, R. and Reder, R., Eds., *Nations and Households in Economic Growth: Essays in Honor of Moses Abramovitz.* New York: Academic Press. www.scirp.org/(S(i43dy n45teexjx455qlt3d2q))/reference/ReferencesPapers.aspx?Referenc eID=1208341

Gauri, Pratik and Van Eerden, Jim, May 16, 2019. "What the Fifth Industrial Revolution is and Why it Matters." *World Economic Forum.* https://europeansting.com/2019/05/16/what-the-fifth-industrial-revolution-is-and-why-it-matters/

Helliwell, John F. et al., 2020. *World Happiness Report 2020.* New York: Sustainable Development Solutions Network. https://worldhappiness.report/ed/2020/

Human Development Report. 2020. "The Next Frontier: Human Development and the Anthropocene." *United Nations Development Programme.* http://hdr.undp.org/en/2020-report

Leitner, Jeff, July 20, 2017. "Where to Start with the SDGs?" *OECD Development Matters.* https://oecd-development-matters.org/2017/07/20/where-to-start-with-the-sdgs/

Marsalis, Wynton, July 11, 2008. "The Cause is People." Interview by *NBC News.* www.nbcnews.com/id/wbna25644088

Mencken, H.L. *The Meaning of Progress.* https://web.stanford.edu/~moore/Chapter1.pdf

New America. OECD. www.sdgsinorder.org/goals

Robert F. Kennedy, March 18, 1968. Remarks at the University of Kansas. https: jfklibrary.org

Schwab, Klaus, January 14, 2016. "The Fourth Industrial Revolution: What It Means, How to Respond." *World Economic Forum.* www.weforum.org/agenda/2016/01/the-fourth-industrial-revolution-what-it-means-and-how-to-respond

Simon, Herbert A. 1969. *The Sciences of the Artificial.* Cambridge: MIT Press.

September 2015. "17 Sustainable Development Goals. 17 Partnerships." *United Nations Sustainable Development Summit.* https://sustainabledevelopment.un.org/content/documents/211617%20Goals%2017%20Partnerships.pdf

CHAPTER 8

Butcher, Mark, April 22, 2020. "When It Comes to Global Governance, Should NGOs Be Inside or Outside the Tent?" *E-International Relations.* www.e-ir.info/2020/04/22/when-it-comes-to-global-governance-should-ngos-be-inside-or-outside-the-tent/

August 19, 2019. "Business Roundtable Redefines the Purpose of a Corporation to Promote 'An Economy That Serves All Americans.'" *Business Roundtable.* www.businessroundtable.org/business-roundtable-redefines-the-purpose-of-a-corporation-to-promote-an-economy-that-serves-all-americans

Cook, Tim, January 28, 2021. Interviewed by Michael Grothaus in "Tim Cook: Privacy and Climate Change are "The Top Issues of the Century." *Fast Company.* www.fastcompany.com/90599049/tim-cook-interview-privacy-legislation-extremism-big-tech

Coster, Helen, September 1, 2021. " 'Oh, That's an Idea…': U.S. Parents Respond to China Screen Time Ban." *Reuters*. www.reuters. com/world/china/oh-thats-an-idea-us-parents-respond-china-screen-time-ban-2021-08-31/

Drennen, Ari and Harding, Sally, March 30, 2021. "Climate Deniers in the 117th Congress." *Center for American Progress*. www.americanprogress.org/article/climate-deniers-117th-congress/

Dunagan, Alana, February 22, 2017. "College Transformed: Five Institutions Leading the Charge in Innovation." *Christensen Institute*. www.christenseninstitute.org/publications/college-transformed/

Fink, Larry. 2019. "Profit & Purpose" *Letter to CEOs*. www.blackrock. com/americas-offshore/en/2019-larry-fink-ceo-letter

Hayhoe, Katharine, December 29, 2021. Interviewed by David Marchese in "An Evangelical Climate Scientist Wonders What Went Wrong." *The New York Times*. www.nytimes.com/interactive/2022/01/03/magazine/katharine-hayhoe-interview.html

"Global Leadership Challenge." *Global Leadership Challenge*. www.leadership-challenge.org/about#:~:text=The%20Global%20Leadership%20Challenge%20%28GLC%29%20aims%20to%20help,fulfil%20personal%20ambition%20but%20furthers%20societies%E2%80%99%

Inglehart, Ronald F. 1977. *The Silent Revolution: Changing Values and Political Styles Am.ong Western Publics*. Princeton University Press.

Inglehart, Ronald F. 1989. *Culture Shift in Advanced Industrial Society*. Princeton University Press.

Kramer, Mark et al., January 16, 2020. "How Global Leaders Should Think About Solving Our Biggest Problems." *Harvard Business Review*. https://hbr.org/2020/01/how-global-leaders-should-think-about-solving-our-biggest-problems

Lopez-Claros, Augusto et al. 2020. *Global Governance and the Emergence of Global Institutions for the 21st Century*. Cambridge University Press.

May 13, 1954. Address by Secretary-General Dag Hammarskjöld at University of California Convocation, Berkeley, California. https://ask.un.org/faq/14623

May 12, 2021. "Covid: Serious Failures in WHO and Global Response, Report Finds." *BBC News*. www.bbc.com/news/world-57085505

Nietzsche, Friedrich. 1895. *The Anti-Christ*, Section 2.

Ord, Toby. 2020. *The Precipice: Existential Risk and the Future of Humanity*. New York: Hachette Books.

2009. "Psychological Factors Help Explain Slow Reaction to Global Warming, Says AOA Task Force." *American Psychological Association*. www.apa.org/news/press/releases/2009/08/climate-change

Slaughter, Anne-Marie, November 12, 2021. "It's Time to Get Honest About the Biden Doctrine." *The New York Times*. www.nytimes.com/2021/11/12/opinion/biden-foreign-policy.html

Smith, Adam. 1759. *The Theory of Moral Sentiments*.

CHAPTER 9

Azhar, Azeem. 2021. *Exponential: How Accelerating Technology is Leaving Us Behind and What to Do About It*. New York: Penguin.

2019. "Amazon will invest over $1.2 billion to provide upskilling training programs for employees." *About Amazon*. www.aboutamazon.com/news/workplace/upskilling-2025

August 2020. "COVID-19: Are children able to continue learning during school closures?" *UNESCO, UNICEF and the World Bank*. https://data.unicef.org/resources/remote-learning-reachability-factsheet/

Bellah, Robert N. et al. 1991. *The Good Society*. New York: Knopf.

Collins, Jim. 1994. *Built to Last: Successful Habits of Visionary Companies*. Harper Business.

Delors, Jacques. 1996. "The Delors Report: Learning: The Treasure Within, the Report to Unesco." *The Delors Commission*. https://db0nus869y26v.cloudfront.net/en/Delors_Report

Draxler, Alexandra. 2013. "A Humanistic Education." Chapter 7 in *Achieving Quality Education for All: A Humanistic Education*. Springer.

Dudley, William C., March 26, 2019. "The Importance of Incentives in Ensuring a Resilient and Robust Financial System." *Speech to the U.S. Chamber of Commerce*, Washington, D.C. Federal Reserve Bank of New York. www.newyorkfed.org/newsevents/speeches/2018/dud180326

Dunagan, Alana, February 22, 2017. "College Transformed: Five Institutions Leading the Charge to Innovation." *Christensen Institute*. www.christensinstitute.org/publicationa/college-transformed/

Finley, Ashley. 2013. "How College Contributes to Workforce Success: Employer Views on What Matters Most." *Association of American Colleges and Universities*. www.aacu.org/research/how-college-contributes-to-workforce-success

"Global Wealth Databook 2022." *Credit Suisse Research Institute*. www.credit-suisse.com/media/assets/corporate/docs/about-us/research/publications/global-wealth-databook-2022.pdf

Jungalwala, Julie, February 19, 2021. "Reinvention Mandate: Succeeding in 2020 and Beyond." *Forbes*. www.forbes.com/sites/forbescoaches council/2021/02/19/reinvention-mandate-succeeding-in-2020-and--beyond/?sh=244089e66d6e

Kegan, Robert. 1982. *The Evolving Self*. Harvard University Press.

King, Martin Luther, Jr., March 31, 1968. "Remaining Awake Through a Great Revolution." *Speech given at Washington D.C.'s National Cathedral*. www.learnoutloud.com/content/blog/archives/2017/04/35_speeches_by_MLK.php

Matthews, Simon C. and Provonost, Peter J., March 2, 2011. "The Need for Systems Integration in Health Care". *JAMA*, 305:9. http://thehtf.org/documents/The_Need_for_Systems_Integration_in_Health_Care.pdf

Mintz, Steven, February 6, 2019. "Higher Education Needs to Innovate. But How?" *Inside Higher Ed*. www.insidehighered.com/blogs/higher-ed-gamma/higher-education-needs-innovate-how

Moore, McKenna, September 30, 2020. "What the Unbanked Need from the 2020 Election." *Fortune*. https://fortune.com/2020/09/30/2020election-trump-biden-financial-inclusion-economic-policy-unbanked-underbanked/

Paine, Thomas. 1796. "Agrarian Justice". *A letter written to the government of France*. www.thomaspaine.org/major-works/agrarian-justice.html

Pettiger, Tejvan, November 25, 2019. "Problems of Capitalism." *Economics Help*. www.economicshelp.org/blog/77/economics/problems-of-capitalism/#:~:text=25%20November%202019%20by%20Tejvan%20Pettinger%20Capitalism%20is,most%20effici ent%20economic%20system%2C%20enabling%20improved%20liv ing%20standards.

Slater, Daniel. "Elements of Amazon's Day 1 Culture." *Amazon Executive Insights.* https://aws.amazon.com/executive-insights/content/how-amazon-defines-and-operationalizes-a-day-1-culture/

Storr, Will. 2019. *The Status Game: On Social Position and How We Use It.* New York: William Collins.

January 22, 2019. "The Education Crisis: Being in School Is Not the Same as Learning." *The World Bank.* www.worldbank.org/en/news/immersive-story/2019/01/22/pass-or-fail-how-can-the-world-do-its-homework

May 24, 2018. "The General Data Protection Regulation Applies in All Member States from 25 May 2018." *Eur-Lex.* https://eur-lex.eur opa.eu/content/news/general-data-protection-regulation-GDPR-applies-from-25-May-2018.html

Welch, Jack. 2000. "General Electric Annual Report 2000 (Annual Report)." *General Electric*, Fairfield, Connecticut, USA.

Wessel, Maxwell et al., November 1, 2016. "The Problem with Legacy Systems." *Harvard Business Publishing.* https://hbsp.harvard.edu/product/R1611D-PDF-ENG

Williamson, Marianne, July 1, 2019. "Health Care in America." *Awaken.* https://awaken.com/2019/07/health-care-in-america-marianne-williamson/#:~:text=by%20Marianne%20Williamson%3A%20The%20biggest%20problem%20with%20America%E2%80%99s,act ual%20cultivation%20of%20health%20and%20prevention%20of%20disease.

December 7, 2019 "What is M-Pesa & How it Works." https://mpesaguide.com/what-is-mpesa/

Vincent, James, October 24, 2018. "Tim Cook Warns of 'Data-Industrial Complex' in Call for Comprehensive US Privacy Laws." *The Verge.* www.theverge.com/2018/10/24/18017842/tim-cook-data-privacy-laws-us-speech-brussels

CHAPTER 10

Abrams, Allison, September 6, 2017. Reneé Carr quoted in "The Psychology Behind Racism." *Psychology Today.* www.psychologytoday.com/us/blog/nurturing-self-compassion/201709/the-psychology-behind-racism

Bakewell, Sarah, March 28, 2023. *Humanly Possible: Seven Hundred Years of Humanist Freethinking, Inquiry, and Hope.* Penguin Press.

Baldwin, James. 1963. *The Fire Next Time.* New York: Dial Press.

Benyus, Janine. 1996. *Biomimicry: Innovation Inspired by Nature.* New York: Harper Perennial.

Cook, Tim, June 19, 2019. "2019 Commencement Address: If You've Built a Chaos Factory You Can't Dodge Responsibility for the Chaos." https://news.stanford.edu/2019/06/16/remrks-tim-cook-2019-commencement

Crutzen, Paul J. and Stoermer, Eugene F. 2000. *The Future of Nature.* Yale University Press.

2022. "Declaration of Modern Humanism." *Humanists International.* https://humanists.international/policy/declaration-of-modern-humanism

Emerson, Ralph Waldo. 1841. *Self-Reliance.*

July, 2009. "Exclusive Interview – 'I am a supporter of globalization.'" Interview of the Dalai Lama. www.dalailama.com/news/2009/exclusive-interview-i-am-a-supporter-of-globalization

Frankl, Victor E. 2014. *Man's Search for Meaning.* Beacon Press.

Friedman, Thomas L. 2016. *Thank You for Being Late: An Optimist's Guide to Thriving in the Age of Accelerations.* New York: Farrar, Straus and Giroux.

Graeber, David, Wengrow, David et al. 2021. *The Dawn of Everything: A New History of Humanity.* Farrar, Straus and Giroux.

Grove, Andrew S. 1983. *High Output Management.* New York: Vintage.

Harari, Yuval N. 2015. *Sapiens: A Brief History of Humankind.* New York: Harper.

Hayhoe, Katharine, December 29, 2021. Interviewed by David Marchese in "An Evangelical Climate Scientist Wonders What Went Wrong." *The New York Times.*

Kornfield, Jack, September 4, 2018, "Our Crisis of Heart: How Compassion Can Strengthen Our Emotional Responses, Our Minds—Our Tech." *One Zero.* https://onezero.medium.com/were-in-a-crisis-of-the-heart-70852fa48b49

Lewis, Ralph, November 25, 2020. "Where Does Morality Come From?" *Psychology Today.* www.psychologytoday.com/intl/blog/finding-purpose/202011/where-does-morality-come

Lightman, Alan, January 15, 2022. "This is No Way to Be Human." *The Atlantic.* www.theatlantic.com/technology/archive/2022/01/machine-garden-natureless-world/621268/

Lloyd, William Forster. 1833. *The Tragedy of the Commons*.

Machiavelli, Niccolò. 1532. *The Prince*.

Marquis, Christopher. 2020. *Better Business: How the B Corp Movement is Remaking Capitalism*. Yale University Press.

Meyer, Robinson, April 16, 2019. "The Cataclysmic Break That (Maybe) Occurred in 1950." *The Atlantic*. www.theatlantic.com/science/archive/2019/04/great-debate-over-when-anthropocene-started/587194/

Miller, John P. 2018. *Love and Compassion: Exploring Their Role in Education*. University of Toronto Press.

Nisbet, Elizabeth and Zelemski, John M, January 2023. "Nature Relatedness and Subjective Well-Being." *Encyclopedia of Quality of Life and Well-Being Research*. https://link.springer.com/referenceworkentry/10.1007/978-3-319-69909-73909-2

O'Leary, Michael and Valdmanis, Warren, March 4, 2021. "An ESG Reckoning is Coming." *Harvard Business Review*. https://hbr.org/2021/03/an-esg-reckoning-is-coming

Segal, Jerome M. 2003. *Graceful Simplicity: The Philosophy and Politics of the Alternative American Dream*. University of California Press.

Soros, George, December 2003. "The Bubble of American Supremacy." *The Atlantic*. www.theatlantic.com/magazine/archive/2003/12/the-bubble-of-american-supremacy/302851/

Roy, Arundhati. 2017. *The Ministry of Utmost Happiness*. New York: Alfred A. Knopf.

Paul, Weiss et al., January 15, 2021. Loretta Lynch quoted in "Sustainability & ESG Year in Review: Key Takeaways". *Lexology*. www.lexicology.com/library/detail.aspx?g=f7cc7642-0acd-413b-87a3-f1c754a3905a

Pink, Daniel. 2011. *Drive: The Surprising Truth About What Motivates Us*. Riverhead Books.

Turkle, Sherry. 2012. *Alone Together: Why We Expect More from Technology and Less from Each Other*. Basic Books.

Visram, Talib, October 8, 2020. "How the B Corporation Movement is Remaking Business." *Fast Company*. www.fastcompany.com/90560758/how-the-b-corp-movement-is-remaking-business?position=2&campaign_date=01282021

von Humboldt, Alexander, 1800. *Kosmos*.

EPILOGUE

Alonzo L. Plough, ed. 2020. *Well-Being: Expanding the Definition of Progress.* The Robert Woods Johnson Foundation.

Solzhenitsyn, Alexandr, June 8, 1978. "A World Split Apart." *Graduation address to Harvard University.* www.americanrhetoric.com/speeches/alexandersolzhenitsynharvard.htm

Index

5G telecommunications, 12
10 Breakthrough Technologies 2023, 23
21st century
   addiction, 12–14
   global society, 115
   humanist code, 192–194

A

Access, right to, 168
Adaptation, innovation as, 180–182
Adoption, gravity of, 39–40
Age of Affluence, The, 77, 101
Age of Commerce, The, 76–77, 101
Age of Conquests, The, 76, 84, 100, 106
Age of Decadence, The, 77–78, 82, 84, 86,
   91, 100, 156
Age of Intellect, The, 77
Age of Mergers and Acquisitions, The, 76
Age of Pioneers, The, 75–76, 84, 100,
   106
Altman, Sam, 28
Amazon, 19
Amazon Web Services (AWS), 19, 150
Amendments to Constitution of America
   First Amendment, 81
   Freedom of Speech amendment, 80
   impact of technology on, 80–81
   Second Amendment, 81
American Dream, 77, 84
American education, 29, 90
American health care system, 147
American Psychological Association
   (APA), 131
American Society of Civil Engineers, 88
Animals of little significance, 189

Annenberg Constitution Day Civics
   Survey, 45
Anthropocene, 184–185, 191
Anthropocene Working Group, 184
Application Program Interface (API), 159
Applin, Sally, 66
Artificial Intelligence (AI), 1–2, 10, 12, 23,
   28, 41, 58, 103
   AI Safety and Security Board (US), 65
   application of, 64, 68
   ChatGPT, 28, 64
   Cyber Challenge, 66
   cybersecurity program for developing
     tools for, 66
   Deep Learning and Machine Learning
     capacities, 64
   "ethical AI" principles, 65
   fraud and deception related to, 66
   as friend or foe, 63–68
   guidelines and guardrails for responsible,
     66
   industrial revolution, 64
   job losses from AI-induced automation
     and robotics, 64
   military use of, 66
   National Security Memorandum for
     directing actions on, 66
   OECD AI ethics guidelines document,
     67
   psychological vulnerabilities, 66
   red-team safety tests, 65
   risks associated with, 65
   safety tests, 65
Atlantic, The (2019), 184
Atomic Bomb Hypocenter Park, in
   Nagasaki, Japan, 82

Printed in the United States
by Baker & Taylor Publisher Services